Survival Guide for General Chemistry

With Math Review and Proficiency Questions

Second Edition

Charles H. Atwood
University of Georgia

with
Carrie G. Shepler

BROOKS/COLE
CENGAGE Learning™

Australia • Brazil • Japan • Korea • Mexico • Singapore • Spain • United Kingdom • United States

ISBN-13: 978-0-495-38751-0
ISBN-10: 0-495-38751-7

Brooks/Cole
20 Davis Drive
Belmont, CA 94002-3098
USA

Cengage Learning is a leading provider of customized learning solutions with office locations around the globe, including Singapore, the United Kingdom, Australia, Mexico, Brazil, and Japan. Locate your local office at: **www.cengage.com/global.**

Cengage Learning products are represented in Canada by Nelson Education, Ltd.

To learn more about Brooks/Cole, visit **www.cengage.com/brookscole.**

Purchase any of our products at your local college store or at our preferred online store **www.ichapters.com.**

Printed in the United States of America
7 8 13 12

Table of Contents

Module 4
The Mole Concept

Module 5
Chemical Reaction Stoichiometry

Module 6
Types of Chemical Reactions

Module 7
Electronic Structure of Atoms

Module 8
Chemical Periodicity

Module 9
Chemical Bonding

Module 10
Molecular Shapes

Module 11
Hybridization and Polarity of Molecules

Module 12
Acids and Bases

Module 13
States of Matter

Module 14
Solutions

Module 15
Heat Transfer, Calorimetry, and Thermodynamics

Module 16
Chemical Kinetics

Module 17
Gas Phase Equilibria

Module 18
Aqueous Equilibria

Module 19
Electrochemistry

Module 20
Nuclear Chemistry

Practice Test Six

Math Review

Preface

To The Student:

In the second edition of the **Survival Guide for General Chemistry with Math Review** we have tried to take the first edition's basic premise and add to it another layer of help for you the student. In the first edition, I tried to write it as if you were sitting beside me at my desk and I was helping you solve problems. Consequently, you still see in this edition solved problems with numerous arrows and boxes pointing where numbers, equations, and other pertinent information come from and are introduced into the problems. All of the most commonly asked questions in a typical general chemistry course and the essence of such a course has been distilled into a small set of topics and included in this edition. Also left from the first edition are the **INSIGHT** and **CAUTION** boxes which should draw your attention to important details that will help you succeed. In this same vein, we have now included a new feature **TIPS** boxes. This feature collects several ideas pertaining to that topic which will help you distinguish that problem from other similar problems and allow you to categorize each problem in a fashion that will improve your success in your classroom. To further assist you in your understanding of chemistry, we have included **Module Predictor Questions**, with worked out solutions, and **Practice Tests** designed to test your knowledge over several Modules. Over the last two years my graduate student, Ms. Kimberly Schurmeier, has analyzed, using Item Response Theory, the tests we have given at the University of Georgia for the last seven years. From her work, we have determined which test questions demonstrate basic, mid-level, and high proficiency understanding of the topics in each Module. As she assisted me with the writing of this second edition, one of the University's Franklin Fellows, Dr. Carrie Shepler, realized that starting each Module with several of these questions would allow you to assess how much you needed to work on that Module. This was the genesis of the **Module Predictor Questions**. Each question in the **Module Predictor Questions** has the proficiency level indicated (1 = basic, 2 = mid-level, and 3 = high) next to it. We suggest that you attempt the **Module Predictor Questions** before reading the Module. If you cannot work the level 2 and 3 questions, it is best that you work on that Module to improve your understanding of those topics. Approximately every five modules, Dr. Shepler included a **Practice Test** with predictor questions which can give you some measure of how well you comprehend the previous modules. Hopefully, this will assist you in studying for tests at your institution. We hope that these new additions will aid your comprehension and success in general chemistry.

Acknowledgements:

Once again, my Brooks/Cole-Thomson colleague, David Harris, has been instrumental in the revision of this **Survival Guide**. His advice and guidance have insured that we improved upon the initial guide and, hopefully, made it even more usable for the students. We are indeed fortunate to have such a friend and colleague.

Kimberly Schurmeier's work in analyzing test questions forms the basis of the **Module Predictor Questions**, a key component of this new edition. We are deeply indebted to Kim for sharing her work with us and helping us determine the question levels. Her research in Item Response Theory analysis of general chemistry questions is cutting edge chemical education research.

Carrie Shepler did the lion's share of this revision. She assembled the **Module Predictor Questions**, wrote the **Practice Tests**, and revised sections of the Modules. Her work on this project along with her continued analysis of students' thought processes on test questions is leading us into new research areas in chemical education. Her work on this revision is genuinely appreciated. She wishes to give many thanks to her husband, Ben, for his love and patience.

My wife, Judy, gets to see me disappear upstairs to my home office nearly every evening to work on this and other projects. She never ceases to amaze me with her love for me and her willingness to let me work on various projects at the expense of her time with me. I cannot adequately express in words my love and appreciation to her.

I dedicate this revision of the Survival Guide to my two children, Louis and Lesley. Both of you have grown up in our house, graduated from high school and college, and then headed off to tackle the world on your own terms. I hope that you continue to grow and to experience the wonderful lives you have set out upon. I love you both.

Module 1 Predictor Questions

The following questions may help you to determine to what extent you need to study this module. The questions are ranked according to ability.

    Level 1 = basic proficiency
    Level 2 = mid level proficiency
    Level 3 = high proficiency

If you can correctly answer the Level 3 questions, then you probably do not need to spend much time with this module. If you are only able to answer the Level 1 problems, then you should review the topics covered in this module.

Level 2   1. How many Mm are there in 427 miles?

Level 2   2. Convert 1.52×10^4 cm^3 to ft^3.

Level 2   3. Determine the number of significant digits in each of the following numbers.
          a) 3700
          b) 770.
          c) 770.0
          d) 0.00420
          e) 8.12×10^4

Level 1   4. Answer the addition problem using the correct number of significant figures: $101.22 + 222.3 =$

Level 1   5. Answer the multiplication problem using the correct number of significant figures: $8.4 \times 8.22 =$

Level 1   6. What is the answer to the numerical calculation, using the correct number of significant digits? $(67.888 - 7.64 + (1.2 \times 10^2)) / 3.27 =$

Level 1   7. A cubic sample of iron has an edge length of 6.32 in. The density of iron is 7.86 g/cm^3. What is the mass of this iron sample?

Module 1 Predictor Question Solutions

1. $\left(\dfrac{427\,\text{mi}}{1}\right)\left(\dfrac{5280\,\text{ft}}{1\,\text{mi}}\right)\left(\dfrac{12\,\text{in}}{1\,\text{ft}}\right)\left(\dfrac{2.54\text{cm}}{1\,\text{in}}\right)\left(\dfrac{10^{-2}\,\text{m}}{1\,\text{cm}}\right)\left(\dfrac{1\,\text{Mm}}{10^{6}\,\text{m}}\right) = 0.687\,\text{Mm}$

2. $\left(\dfrac{1.52\times10^{4}\,\text{cm}^{3}}{1}\right)\left(\dfrac{(1\,\text{in})^{3}}{(2.54\,\text{cm})^{3}}\right)\left(\dfrac{(1\,\text{ft})^{3}}{(12\,\text{in})^{3}}\right) = 0.537\,\text{ft}^{3}$

3. a) 2 significant figures; the trailing zeros are not significant
   b) 3 significant figures; the trailing zero is significant due to the decimal
   c) 4 significant figures; the trailing zeros are significant due to the decimal
   d) 3 significant figures; only the ending zero is significant
   e) 3 significant figures; the digits in 10^{4} are not significant

4.
$$\begin{array}{r} 101.22 \\ +\,222.3 \\ \hline 323.5 \end{array}$$
   The answer is limited to one decimal place.

5. The answer is limited to two significant figures; 69

6. The subtraction step is limited to two decimal places (four significant figures). The addition step is limited to one decimal place (two significant figures) because the first place that two numbers have a significant figure in common is in the tens place. The division step is then limited to two significant figures. The answer is 55.

7. $(6.32\,\text{in})^{3} = 2.52\times10^{2}\,\text{in}^{3}\left(\dfrac{(2.54\,\text{cm})^{3}}{(1\,\text{in})^{3}}\right) = 4.14\times10^{3}\,\text{cm}^{3}$

   $\left(\dfrac{7.86\,\text{g}}{\text{cm}^{3}}\right)\left(\dfrac{4.14\times10^{3}\,\text{cm}^{3}}{1}\right) = 3.25\times10^{4}\,\text{g}$

Module 1
Metric System, Significant Figures,
Dimensional Analysis, and Density

Introduction
This module addresses several topics that are typically introduced in the first chapter of general chemistry textbooks. This module describes:

1. the basic rules of the metric system and significant figures
2. how to use dimensional analysis to solve problems
3. the relationship between density, mass, and volume and how to apply dimensional analysis to density problems

---

Module 1 Key Equations & Concepts

1. $\text{density} = \dfrac{\text{mass}}{\text{volume}} = \dfrac{m}{v}$

   If any two of the variables in the equation are known, then you can solve for the third using basic algebra.

The metric system uses a series of multipliers to convert from one sized unit to another size. You must be very familiar with these prefixes and how to convert from one size unit to another. A common set of multiplier prefixes is given in the table below.

It is important to recognize that these prefixes may be used with any unit of measurement, and that the relationship between the *base unit* and the unit with the prefix is always the same regardless of the base unit. The base unit is represented by *x* in the table.

Pay special attention to the unit factors provided as they are what will be used in converting one unit to another. Note that each unit factor may be written in two equivalent ways. The one you use depends on what units you are trying to cancel in a dimensional analysis problem (see examples below).

One way to help insure that you work conversion problems correctly is to remember which one of the units is the largest. For example, if you are converting from pg to Mg, then keep in mind that a Mg is much, much larger than a pg. So, the numerical value should get much smaller as you convert from pg to Mg

Prefix Name	Prefix Symbol	Multiplication Factor	Unit Factors
mega-	M	1000000 or 10^6	$\dfrac{1\,Mx}{10^6\,x} = \dfrac{10^6\,x}{1\,Mx}$
kilo-	k	1000 or 10^3	$\dfrac{1\,kx}{10^3\,x} = \dfrac{10^3\,x}{1\,kx}$
deci-	d	0.1 or 10^{-1}	$\dfrac{1\,dx}{10^{-1}\,x} = \dfrac{10^{-1}\,x}{1\,dx}$
centi-	c	0.01 or 10^{-2}	$\dfrac{1\,cx}{10^{-2}\,x} = \dfrac{10^{-2}\,x}{1\,cx}$
milli-	m	0.001 or 10^{-3}	$\dfrac{1\,mx}{10^{-3}\,x} = \dfrac{10^{-3}\,x}{1\,mx}$
micro-	μ	0.000001 or 10^{-6}	$\dfrac{1\,\mu x}{10^{-6}\,x} = \dfrac{10^{-6}\,x}{1\,\mu x}$
nano-	n	0.000000001 or 10^{-9}	$\dfrac{1\,nx}{10^{-9}\,x} = \dfrac{10^{-9}\,x}{1\,nx}$
pico-	p	0.000000000001 or 10^{-12}	$\dfrac{1\,px}{10^{-12}\,x} = \dfrac{10^{-12}\,x}{1\,px}$

Sample Exercises

1. How many mm are there in 3.45 km?
   The correct answer is 3.45×10^6 mm

The table indicates that there are 1000 m in 1 km and that 1 mm = 0.001 m.

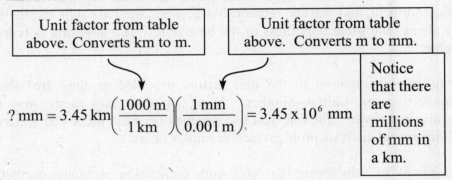

Unit factor from table above. Converts km to m.

Unit factor from table above. Converts m to mm.

$$? \text{ mm} = 3.45 \text{ km}\left(\frac{1000\,m}{1\,km}\right)\left(\frac{1\,mm}{0.001\,m}\right) = 3.45 \times 10^6 \text{ mm}$$

Notice that there are millions of mm in a km.

Note that the km and the m both cancel. The canceling of units is the key to dimensional analysis problems.

In this problem, the km is a much larger unit than the mm. Thus we should expect that there will be many of the smaller unit, mm's, in the large units. The answer 3.45×10^6 mm is sensible.

TIP Always convert to a base unit (like m or g) first. Then proceed to a different unit if necessary.

2. How many mg are there in 15.0 pg?
   The correct answer is 1.5×10^{-8} mg

From the table we see that 1 pg = 10^{-12} g and 1 mg = 10^{-3} g.

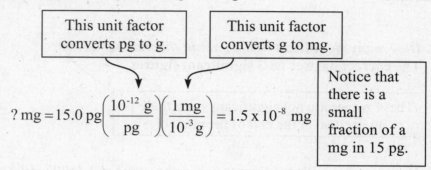

This unit factor converts pg to g.

This unit factor converts g to mg.

Notice that there is a small fraction of a mg in 15 pg.

$$? \, mg = 15.0 \, pg \left(\frac{10^{-12} \, g}{pg} \right) \left(\frac{1 \, mg}{10^{-3} g} \right) = 1.5 \times 10^{-8} \, mg$$

In this problem picograms, pg, are the smaller unit. We should expect that there are very few milligrams, mg, in 15.0 pg. The correct answer is 1.5×10^{-8} mg, which is reasonable.

Significant Figures
All non-zero integers are significant. When determining the number of significant figures in a value, by far the most confusion revolves around zeros. Sometimes they are significant, and sometimes they are not! Below are some rules to help you determine whether or not a zero is significant.

1. Zeros located between two integers ARE significant.
2. Zeros located at the ends of numbers containing decimals ARE significant.
3. Zeros located between an integer on the right and a decimal on the left ARE significant.
4. Zeros used as place holders to indicate the position of a decimal ARE NOT significant. This includes a zero at the end of a number that does not contain a decimal.

Sample Exercises
3. How many significant figures are in the number 58062?
   The correct answer is: 5 significant figures

This zero is significant because it is embedded in other significant digits. See rule 1.

4. How many significant figures are in the number 0.0000543?
   The correct answer is: 3 significant figures

None of these zeroes are significant because their purpose is to indicate the position of the decimal place. Only the non-zero integers are significant.

5

5. *How many significant figures are in the number 0.009120?*
   The correct answer is: 4 significant figures

This zero is significant.
See rule 4!

These three zeroes are not significant because they are place holders.

TIPS

6. *How many significant figures are in the number 24500?*
   The correct answer is: 3 significant figures

These zeroes are not significant since there is
no decimal at the end of the number.

7. *How many significant figures are in the number 2.4500 x 10^4?*
   The correct answer is: 5 significant figures

As written both of these zeroes are significant because the number contains a decimal.

Notice that this number is the same as a previous exercise, but written in scientific notation. All of the same rules apply.

CAUTION

None of the numbers in the 10^x portion of numbers written in scientific notation are significant.

Calculations and Significant Figures
Rules for determining the number of significant figures in the answer to a calculation depend on the mathematical operation being performed.
- In addition and subtraction problems, the final answer must contain no digits beyond the most doubtful digit in the numbers being added or subtracted.
- In multiplication and division problems involving significant figures the final answer must contain the same number of significant figures as the number with the least number of significant figures.

Sample Exercises
8. *What is the sum of 12.674 + 5.3150 + 486.9?*
   The correct answer is: 504.9

This 9 is in the tenths decimal place. It is the most doubtful digit in the sum.

6

The most doubtful digit in each of the numbers is underlined 12.674, 5.3150, 486.9. Notice that the 486.9 has the most doubtful digit because the 9 is only in the tenths position and the other numbers are doubtful in the thousandths and ten thousandths positions. *The final answer must have the final digit in the tenths position.*

9. *What is the correct answer to this problem: $2.6138 \times 10^6 - 7.95 \times 10^{-3}$?*
  The correct answer is: 2.6138×10^6

> This 8 is the most doubtful digit in the sum. It is in the hundreds position.

The number 2.6138×10^6 can be also written as 2,613,800. Its most doubtful digit, the 8, is in the hundreds position. The other number, 7.95×10^{-3}, can be written as 0.00795. Its most doubtful digit, the 5, is in the one millionths position. Consequently, the final answer cannot extend beyond the 8 in 2.6138×10^6.

 TIP When adding and subtracting, both numbers must be expressed to the same power of 10 to determine the most doubtful digit.

10. *What is the correct answer to this problem: 47.893×2.64?*
  The correct answer is: 126

> This number contains only 3 significant digits, so the answer can have only 3 significant figures.

11. *What is the correct answer to this problem: $1.95 \times 10^5 \div 7.643 \times 10^{-4}$?*
  The correct answer is 2.55×10^8

> This number contains 3 significant figures.

> This number contains 4 significant figures.

Just as in exercise 10, the number with fewest significant digits determines that the final answer must also have three significant digits.

Dimensional Analysis

In chemistry we often perform calculations that require changing from one set of units to a second set of units. Dimensional analysis is a convenient method to help convert units without making arithmetic errors. In this method, common conversion factors given in your textbook are arranged so that one set of units cancels, converting the problem to the second set of units.

12. How many Mm are in 653 ft?
   The correct answer is: 1.99 x 10⁻⁴ Mm.

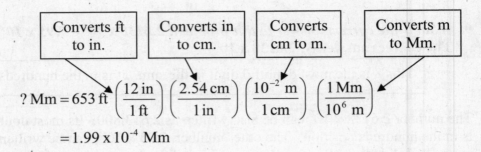

Notice that the problem is arranged so that each successive conversion factor makes progress in the conversion process. Feet are converted to inches, then to cm, next to m, and finally to Mm. This is the simplest kind of dimensional analysis problem because all of the units are linear.

13. How many km² are in 2.5 x 10⁸ in²?
   The correct answer is: 1.6 x 10⁻¹ km²

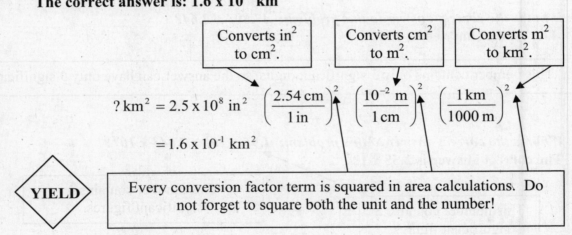

YIELD

Every conversion factor term is squared in area calculations. Do not forget to square both the unit and the number!

Because the problem involves area (a two dimensional unit) all of the conversion factors are similar to exercise 12, but they must be squared to be in the appropriate units.

TIP

The ( )² notation around a unit factor literally means that you are multiplying the unit factor by itself. If you have trouble remembering to square the unit factor, then try writing it out as the unit factor multiplied by the unit factor.

14. How many yd³ are in 7.93 x 10¹² cm³?
   The correct answer is: 1.04×10^7 yd³

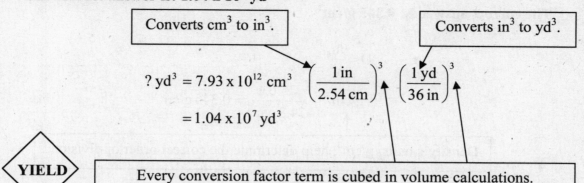

Converts cm³ to in³.

Converts in³ to yd³.

$$? \, \text{yd}^3 = 7.93 \times 10^{12} \, \text{cm}^3 \left(\frac{1 \, \text{in}}{2.54 \, \text{cm}} \right)^3 \left(\frac{1 \, \text{yd}}{36 \, \text{in}} \right)^3$$

$$= 1.04 \times 10^7 \, \text{yd}^3$$

YIELD — Every conversion factor term is cubed in volume calculations.

Density

15. What is the mass, in g, of a 68.2 cm³ sample of ethyl alcohol? The density of ethyl alcohol is 0.789 g/cm³.
   The correct answer is: 53.8 g

Density converts the volume of a substance into the mass.

$$D = \frac{m}{V} \Rightarrow m = DV$$

$$? \, \text{g} = 68.2 \, \text{cm}^3 \left(\frac{0.789 \, \text{g}}{1 \, \text{cm}^3} \right)$$

$$= 53.8 \, \text{g}$$

The final units are g because the cm³ in the density cancels with the original volume.

16. What is the volume, in cm³, of a 237.0 g sample of copper? The density of copper is 8.92 g/cm³.
   The correct answer is: 26.6. cm³

$$D = \frac{m}{V} \Rightarrow V = \frac{m}{D}$$

$$? \, \text{cm}^3 = 237.0 \, \text{g} \left(\frac{1 \, \text{cm}^3}{8.92 \, \text{g}} \right)$$

$$= 26.6 \, \text{cm}^3$$

17. **What is the density of a substance having a mass of 25.6 g and a volume of 74.3 cm³?**

    The correct answer is: 0.345 g/cm³

$$D = \frac{m}{V}$$

$$? \, g/cm^3 = \frac{25.6 \, g}{74.3 \, cm^3} = 0.345 \, g/cm^3$$

Density's units, g/cm³, help determine the correct order of division.

Module 2 Predictor Questions

The following questions may help you to determine to what extent you need to study this module. The questions are ranked according to ability.

Level 1 = basic proficiency
Level 2 = mid level proficiency
Level 3 = high proficiency

If you can correctly answer the Level 3 questions, then you probably do not need to spend much time with this module. If you are only able to answer the Level 1 problems, then you should review the topics covered in this module.

Level 1   1. How many atoms of each element are in one molecule or formula unit of each of the following?
a) C_4H_9OH
b) $MgBr_2$
c) $Ba_3(PO_4)_2$

Level 1   2. How many ions are present in one formula unit of each of the following?
a) NaCl
b) $BaCl_2$
c) $NaNO_3$
d) $Al(NO_3)_3$
e) $Al_2(CO_3)_3$

Level 3   3. How many aluminum, phosphate, phosphide, and oxide ions are in one formula unit of $AlPO_4$?

Level 1   4. How many atoms of P are in one mole of $Mg_3(PO_4)_2$?

Module 2 Predictor Question Solutions

1. a) There are: four C atoms
ten H atoms
one O atom

   b) There are: one Mg atom
two Br atoms

   c) There are: three Ba atoms
two P atoms
eight O atoms

2. a) two ions (Na^+ and Cl^-)
   b) three ions (Ba^{2+} and two Cl^-)
   c) two ions (Na^+ and NO_3^-)
   d) four ions (Al^{3+} and three NO_3^-)
   e) five ions (two Al^{3+} and three CO_3^{2-})

3. There is one Al^{3+} ion and one PO_4^{3-} ion. The phosphate ion is a polyatomic ion that does not break down into further ions, so there are no phosphide or oxide ions present.

4. $\left(1 \, mol \, Mg_3(PO_4)_2\right)\left(\dfrac{2 \, mol \, P}{1 \, mol \, Mg_3(PO_4)_2}\right)\left(\dfrac{6.022 \times 10^{23} \, P \, atoms}{1 \, mol \, P}\right) = 1.20 \times 10^{24} \, P \, atoms$

Module 2
Understanding Chemical Formulas

Introduction

What information is contained in a chemical formula and how do we interpret that information? Chemists use specific symbolism to express their understanding of elements, compounds, ions and ionic compounds. The primary goal of this module is to help you:

1. recognize these symbols
2. learn how to determine the number and types of atoms or ions present in a substance.

Module 2 Key Equations & Concepts

1. *Molecular formulas*
   Indicate the number of each **atom** present in a **molecule** (C_5H_{12})
2. *Ionic formulas*
   Indicate the number of each **ion** present in a **formula unit** ($Al_2(CO_3)_3$)
   Also indicate the number of each atom present
3. *Stoichiometric coefficients*
   Indicate the number of a particular molecule or formula unit in the chemical symbolism; found in balanced chemical equations

---

Sample Exercises
Interpreting Chemical Formulas

1. **How many atoms of each element are present in one molecule of C_2H_5OH?**
   The correct answer is: 2 C, 6 H, and 1 O

$$C_2H_5OH$$

There are 2 carbon atoms, 1 oxygen atom, and 6 hydrogen atoms in one molecule. The molecular formula displays the number of atoms in each molecule of a species.

Do not forget that if there is no subscript written, it is understood that there is one atom of that element present.

2. **How many atoms of each element are present in one formula unit of $Al_2(SO_4)_3$?**
   The correct answer is: 2 Al, 3 S, and 12 O

$$Al_2(SO_4)_3$$

There are 2 aluminum atoms, 3 sulfur atoms, and 12 oxygen atoms.
Remember, numbers outside a parenthesis are multiplied times the subscripts of all the elements inside the parentheses. Thus there are 3 x 1 = 3 sulfur atoms and 3 x 4 = 12 oxygen atoms.

Note that this is an example of an *ionic compound* (see Module 3). The parentheses around (SO_4) indicate that it is a *polyatomic ion*. Its actual formula is SO_4^{2-}. Two Al^{3+} ions are required to balance the charge of the three SO_4^{2-}. So, this formula also tells us that there are 2 Al^{3+} ions and 3 SO_4^{2-} ions for a total of 5 ions.

Using Stoichiometric Coefficients

3. ***How many atoms of each element are present in 3 molecules of C_5H_{12} ?***
   The correct answer is: 15 C, 36 H

$$3\ C_5H_{12}$$

There are 15 carbon atoms and 36 hydrogen atoms in 3 C_5H_{12}.
3 x 5 = 15 C atoms and 3 x 12 = 36 H atoms

The 3 represents a *stoichiometric coefficient* as you would find in a balanced chemical equation.

4. ***How many atoms of each element are present in five formula units of $Ca_3(PO_4)_2$?***
   How many ions are in one formula unit of $Ca_3(PO_4)_2$?
   The correct answer is: 15 Ca, 10 P, and 40 O in five formula units; 5 ions in one formula unit

$$5\ Ca_3(PO_4)_2$$

There are 15 calcium atoms, 10 phosphorus atoms, and 40 oxygen atoms.
5 x 3 = 15 Ca atoms, 5 x 2 = 10 P atoms, and 5 x 4 x 2 = 40 O atoms
The formula also tells us that there are 3 Ca^{2+} ions and 2 PO_4^{3-} ions in one formula unit.

 Remember that the stoichiometric coefficient is multiplied by *each* subscript.

Interpreting Chemical Formulas

5. ***Using circles to represent the atoms, draw your best representation of what C_4H_{10} would look like if we could see atoms, ions, and molecules.***

$$C_4H_{10}$$

The 4 carbon atoms are in the center of the molecule.

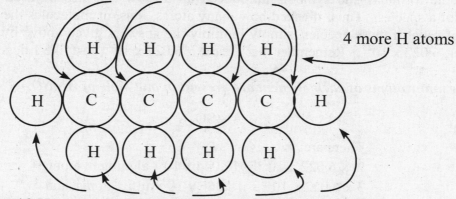

4 more H atoms

The 10 H atoms are around the outside of the molecule.
Notice, that this is one single molecule not 14 separate things.

From a chemical standpoint this is not the only way to draw C_4H_{10}, but all of the possibilities will consist of molecules with the atoms connected.

6. *Using circles to represent the atoms and ions, draw your best representation of what $Sr_3(PO_4)_2$ would look like if we could see atoms, ions, and molecules. Remember, ions are independent species.*

Notice that the three Sr^{2+} ions are independent species.

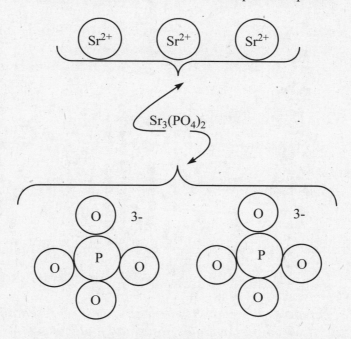

The two PO_4^{3-} ions are also independent species.

Using Chemical Formulas to Determine Numbers of Atoms in One Mole of a Substance

All of the formulas and symbols introduced up to now can also be used to represent moles of a species. Thus, if asked how many atoms, ions, or molecules there are in one mole of each of these species, simply multiply the answers given above by Avogadro's number, 6.022×10^{23}. Remember that 1 mole $= 6.022 \times 10^{23}$, just like 1 dozen $= 12$.

7. How many atoms of each element are present in one mole of $Al_2(SO_4)_3$?

$$Al_2(SO_4)_3$$

There are:

$2 \times 6.022 \times 10^{23} = 12.044 \times 10^{23}$ aluminum atoms,

$3 \times 6.022 \times 10^{23} = 18.066 \times 10^{23}$ sulfur atoms, and

$12 \times 6.022 \times 10^{23} = 72.264 \times 10^{23}$ oxygen atoms.

It is very important to remember that chemical formulas give us lots of different types of information. You must have a good understanding of the difference between atoms, ions, and molecules to correctly apply the different types of information.

Module 3 Predictor Questions

The following questions may help you to determine to what extent you need to study this module. The questions are ranked according to ability.

Level 1 = basic proficiency
>Level 2 = mid level proficiency
>Level 3 = high proficiency

If you can correctly answer the Level 3 questions, then you probably do not need to spend much time with this module. If you are only able to answer the Level 1 problems, then you should review the topics covered in this module.

1. Name the following inorganic compounds.

Level 1	a) N_2O_5
Level 1	b) $MgCl_2$
Level 2	c) $Al(OH)_3$
Level 2	d) $HClO_4$
Level 3	e) $NaClO_3$
Level 3	f) $KHSO_4$

2. Write the formulas of the following inorganic compounds.

Level 1	a) sulfur trioxide
Level 1	b) cesium bromide
Level 2	c) ammonium chloride
Level 2	d) nitric acid
Level 3	e) potassium sulfate
Level 3	f) calcium hydrogen phosphate

3. Determine the number of ions in one formula unit of each of the following compounds.

Level 1	a) CO_2
Level 1	b) H_3PO_4
Level 2	c) $Al(OH)_3$
Level 3	d) KNO_2
Level 3	e) $Al(OH)Cl_2$

Module 3 Predictor Question Solutions

1. a) dinitrogen pentoxide
   b) magnesium chloride
   c) aluminum hydroxide
   d) sodium chlorate
   e) potassium hydrogen sulfate OR potassium bisulfate

2. a) SO_3
   b) $CsBr$
   c) NH_4Cl
   d) HNO_3
   e) K_2SO_4
   f) $CaHPO_4$

3. a) CO_2 is a covalent compound; there are no ions present
   b) H_3PO_4 dissociates into 3 H^+ and 1 PO_4^{3-}; there are four ions
   c) $Al(OH)_3$ dissociates into 1 Al^{3+} and 3 OH^-; there are four ions
   d) KNO_2 dissociates into 1 K^+ and 1 NO_2^-; there are two ions
   e) $Al(OH)Cl_2$ dissociates into 1 Al^{3+}, 1 OH^-, and 2 Cl^-; there are four ions

Module 3
Chemical Nomenclature

Introduction

Chemical nomenclature is the chemist's language. In order to ensure that everything is interpreted correctly, we must follow a specific set of nomenclature rules. This module will:

1. familiarize you with the rules of chemical nomenclature
2. help you to recognize various types of chemical compounds and then to apply the appropriate nomenclature rules

---

Module 3 Key Concepts

Ionic compounds:

Ionic compounds are those composed of a metal cation and some anion. There are several types of ionic compounds, each with their own rules for naming.

Metal cations combined with:

1. **Nonmetal anions** (simple binary ionic compounds)
    Nomenclature is the metal's name followed by nonmetal's stem plus –ide. If the metal cation is a transition metal, then add the oxidation state in parentheses after then metal's name.
2. **Polyatomic anions** (pseudobinary ionic compounds)
    Nomenclature is metal's name followed by polyatomic ion's name.
    Consult your textbook for a list of common polyatomic ions whose names and formulas you should recognize.

Covalent compounds

These are compounds composed of two or more nonmetals.

3. **Two nonmetals** (binary covalent compounds)
    The less electronegative element is named first, and the more electronegative is named second using stem plus –ide. Prefixes such as di-, tri-, etc. are used for both elements.
4. **Hydrogen combined with a nonmetal in aqueous solution** (binary acid)
    Nomenclature is hydro followed by nonmetal stem with suffix –ic acid.
5. **Hydrogen, oxygen, and a nonmetal combined in one compound** (ternary acids)
    Nomenclature is a series of names based upon the oxidation state of the nonmetal.
    Nonmetal highest oxidation state is per stem –ic acid.
    Nonmetal second highest oxidation state is stem –ic acid.
    Nonmetal third highest oxidation state is stem –ous acid.
    Nonmetal lowest oxidation state is hypo stem –ous acid.

> **Special types of ionic compounds**
> 6. **Metal ions combined with a polyatomic ion made from a ternary acid**
>   Ternary acid salts – nomenclature is the metal's name followed by the same series of names used for the ternary acid with two changes. The –ic suffixes are changed to –ate and the –ous suffixes are changed to –ite.
> 7. **Metal ions combined with a ternary acid salt and hydrogen.**
>   Acidic salts of ternary acids – nomenclature is the metal's name followed by hydrogen (including the appropriate di-, tri-, etc. prefix) plus the ternary acid salt's name.
> 8. **Metal ions combined with hydroxyl groups and nonmetal ions.**
>   Basic salts of polyhydroxy bases – nomenclature is the metal's name followed by hydroxy (including the appropriate di-, tri-, etc. prefix) plus the nonmetal stem plus –ide.

Sample Exercises

1. What is the correct name of the chemical compound $CaBr_2$?
   The correct answer is: calcium bromide

Metal cations and nonmetal anions make <u>simple binary ionic compounds</u>. Simple binary ionic compounds are named using the metal's name followed by the nonmetal's stem and the suffix –ide. Prefixes like di- or tri- are ***not used*** to denote the number of ions present in the substance.

The metal cation in this case is Ca^{2+}, the calcium ion. The anion is Br^-, from the element bromine, whose ending is changed to –ide.

2. What is the correct name of the chemical compound $Mg_3(PO_4)_2$?
   The correct answer is: magnesium phosphate

Metal cations and polyatomic anions make <u>pseudobinary ionic compounds</u>. These compounds are named using the metal's name followed by the correct name of the polyatomic anion. Your textbook has a list of the polyatomic anions that you are expected to know. Make sure that you have the name, the anion's formula, and the charge memorized. Once again, no prefixes are used in these compounds to tell the number of ions present. Mg^{2+} is a positive ion made from the metal magnesium. PO_4^{3-} is a negative polyatomic ion named phosphate.

3. What is the correct name of this chemical compound, $FeCl_3$?
   The correct answer is: iron (III) chloride

Transition metal cations and nonmetal or polyatomic anions make <u>transition metal ionic compounds</u>. Their names are derived from the metal's name followed by the metal's oxidation state in Roman numerals inside parentheses. A metal's oxidation state is determined from the oxidation state of the anion. Fe^{3+} is a positive ion made from a transition metal (B Groups on the periodic chart). Cl^{1-} is a negative ion made from the nonmetal chlorine.

4. *What is the correct name of the chemical compound, N_2O_4?*
   The correct answer is: dinitrogen tetroxide

This compound is made from two nonmetals, nitrogen and oxygen, so it is a <u>binary covalent compound</u>. *These compounds use prefixes to indicate the number of atoms of each element present in the compound.* This is an important difference from the ionic compounds in the previous examples.

5. *What is the correct name of the chemical compound $H_2S(aq)$?*
   The correct answer is: hydrosulfuric acid

This compound is made from hydrogen and a nonmetal. Furthermore, the symbol $_{(aq)}$ also indicates that this compound is dissolved in water. That combination is indicative of a <u>binary acid</u>. Binary acids are named using the prefix hydro- followed by the nonmetal's stem and the suffix –ide.

If the symbol (aq) is not present, then the compound is named as a binary covalent compound. In this case H_2S without the $_{(aq)}$ would be named dihydrogen sulfide.

6. *What is the correct name of this chemical compound, $HClO_3$?*
   The correct answer is: chloric acid

This compound is made from three nonmetals, H, O, and another nonmetal, chlorine. This combination of nonmetals is a <u>ternary acid</u>. Ternary acids are named based on a system derived from the third nonmetal's oxidation state. (See your textbook for more information on assigning oxidation numbers.) The easiest method to learn these compounds is to use the table of "ic acids" found in your textbook. You must learn both the compound's formula and its name. Once you know the "ic acids" then use the following system:
   The acid with one more O atom than the "ic acid" is the "per stem ic acid".
   One fewer O atom than the "ic acid" is the "ous acid".
   Two fewer O atoms than the "ic acid" is the "hypo stem ous acid".

7. *What is the correct name of the chemical compound $KClO_4$?*
   The correct answer is: potassium percholorate

This compound is made from a metal ion, K^+, and a polyatomic anion that is derived from the ternary acids discussed above. It is called a <u>ternary acid salt</u>. The anion's name is based upon the ending of the ternary acid. Ternary acids ending in "ic" give salts that end in "ate". Ternary acids that end in "ous" give salts that end in "ite". The prefixes per- and hypo- are retained.

The chloric acid series of potassium salts are shown below:
$KClO_4$ is potassium perchlorate. $KClO_3$ is potassium chlorate. $KClO_2$ is potassium chlorite. Finally, $KClO$ is potassium hypochlorite.

 **CAUTION** These compounds are some of the most difficult to name, so pay special attention to them.

8. What is the correct name of the chemical compound NaH_2PO4?
The correct answer is: sodium dihydrogen phosphate

This compound is made from a metal cation, Na^+, and a polyatomic anion made from a ternary acid that still retains some of its acidic hydrogens. These compounds are called <u>acidic salts of ternary acids</u>. The names for these compounds use the word hydrogen plus a prefix, in this case di-, to indicate the number of acidic hydrogens that are present. The last part of the salt's name is the same as determined in question 7 for the ternary acid salts.

9. What is the correct name of the chemical compound $Al(OH)_2Cl$?
The correct answer is: aluminum dihydroxy chloride

This compound is made from a metal ion, Al^{3+}, and three anions (two hydroxide ions and one chloride ion). Compounds containing hydroxide ions and other anions plus a metal ion are <u>basic salts of polyhydroxy bases</u>. Their names must indicate the number of OH^{1-} groups that are present in the compound. This is done using the appropriate prefix attached to hydroxy. The remainder of the compound's name is the same as for binary ionic compounds.

TIPS

1. If the compound contains a metal cation, then you should NOT use prefixes in the name. Prefixes are used only in covalent compounds.
2. Check the location of the metal on the periodic table. If the metal is a transition metal, then you likely need to use a Roman numeral to indicate its oxidation state (there are a few exceptions; see your textbook)
3. Be sure that you are familiar with the names and formulas of the common polyatomic ions.
4. Pay careful attention to whether a binary acid is written with (g) or (aq). Binary acids have different names depending on those symbols!
5. Be sure you know the names and formulas (including charges) of *either* "-ic acids or the "-ate ions."

22

Practice Test One
Modules 1-3

Level 2   1. How many dm^3 are there in 3.24×10^8 in^3?

Level 2   2. Determine the correct answer to this numerical calculation using the correct number of significant figures. Determine how many significant figures are in each of the numbers.

$$((27.340 - 6.00) \times (6.8371 \times 10^3)) + 871.4$$

Level 1   3. The density of mercury is 13.59 g/cm^3. What is the mass of a sample of mercury with a volume of 0.0230 in^3?

Level 1   4. How many atoms of N are there in 2.57 mol $Al(NO_3)_3$?

Level 1   5. How many moles of $Ca_3(AsO_4)_2$ are in 723.2 g of $Ca_3(AsO_4)_2$?

Level 3   6. How many of each of the following are there in one formula unit of $Fe_2(SO_4)_3$?

        a) iron (III) ions
        b) sulfate ions
        c) sulfide ions
        d) oxide ions

Levels 1-3   7. Name the following compounds.
        a) PCl_5
        b) $(NH_4)_2SO_4$
        c) $LiNO_3$
        d) KH_2BrO
        e) XeF_4

Levels 1-3   8. Write the formulas for the following compounds.
        a) sulfur hexafluoride
        b) hydrocyanic acid
        c) copper (II) monohydroxy chloride
        d) magnesium bromide
        e) hypochlorous acid

Module 4 Predictor Questions

The following questions may help you to determine to what extent you need to study this module. The questions are ranked according to ability.

Level 1 = basic proficiency
    Level 2 = mid level proficiency
    Level 3 = high proficiency

If you can correctly answer the Level 3 questions, then you probably do not need to spend much time with this module. If you are only able to answer the Level 1 problems, then you should review the topics covered in this module.

Level 1   1. Determine the molar mass of C_6H_{14}

Level 1   2. Determine the formula weight of $Mg(ClO_3)_2$

Level 1   3. How many molecules of C_2H_5OH are in 0.342 mols of C_2H_5OH?

Level 2   4. How many nitrate ions are in 0.147 mols of $Zn(NO_3)_2$?

Level 1   5. What is the mass, in grams, of 0.348 mols of C_3H_7OH?

Level 1   6. What is the mass, in grams, of 6.34×10^{22} atoms of Sc?

Level 3   7. What is the total mass, in grams, of the Cl atoms in 0.483 mols of $Ba(ClO_4)_2$?

Level 2   8. Determine the molar mass of the compound: $Al_2(SO_4)_3$
           a) How many mols of $Al_2(SO_4)_3$ are present in 94.2 g of $Al_2(SO_4)_3$?
           b) How many O atoms are present in 37.5 g of $Al_2(SO_4)_3$

Module 4 Predictor Question Solutions

1. C_6H_{14}
   C: 12.01 g/mol x 6 = 72.06 g/mol
   H: 1.01 g/mol x 14 = 14.14 g/mol

   (72.06 g/mol) + (14.14 g/mol) = 86.20 g/mol

2. $Mg(ClO_3)_2$
   Mg = 24.31 g/mol
   Cl = 35.45 g/mol x 2 = 70.90 g/mol
   O = 16.00 g/mol x 6 = 96.00 g/mol

   (24.31 g/mol) + (70.90 g/mol) + (96.00 g/mol) = 191.21 g/mol

3. $(0.342 \, \text{mol} \, C_2H_5OH)\left(\dfrac{6.022 \times 10^{23} \, \text{molecules}}{1 \, \text{mol} \, C_2H_5OH}\right) = 2.06 \times 10^{23} \, C_2H_5OH \, \text{molecules}$

4. $(0.147 \, \text{mol} \, Zn(NO_3)_2)\left(\dfrac{2 \, \text{mol} \, Zn^{2+}}{1 \, \text{mol} \, Zn(NO_3)_2}\right)\left(\dfrac{6.022 \times 10^{23} \, Zn^{2+}}{1 \, \text{mol} \, Zn^{2+}}\right) = 1.77 \times 10^{23} \, Zn^{2+} \, \text{ions}$

5. $(0.348 \, \text{mol} \, C_3H_7OH)\left(\dfrac{60.11 \, \text{g} \, C_3H_7OH}{1 \, \text{mol} \, C_3H_7OH}\right) = 20.9 \, \text{g} \, C_3H_7OH$

6. $(6.34 \times 10^{22} \, \text{atoms Sc})\left(\dfrac{1 \, \text{mol Sc}}{6.022 \times 10^{23} \, \text{atoms Sc}}\right)\left(\dfrac{44.96 \, \text{g Sc}}{1 \, \text{mol Sc}}\right) = 4.73 \, \text{g Sc}$

7. $(0.483 \, \text{mol} \, Ba(ClO_4)_2)\left(\dfrac{2 \, \text{mol Cl atoms}}{Ba(ClO_4)_2}\right)\left(\dfrac{35.45 \, \text{g Cl}}{1 \, \text{mol Cl atoms}}\right) = 34.24 \, \text{g Cl}$

8. $Al_2(SO_4)_3$
   Al = 26.98 g/mol x 2 = 53.96 g/mol
   S = 32.07 g/mol x 3 = 96.21 g/mol
   O = 16.00 g/mol x 12 = 192.00 g/mol

   Formula weight = 342.17 g/mol

a) $(94.2 \text{ g Al}_2(\text{SO}_4)_3)\left(\dfrac{1 \text{ mol Al}_2(\text{SO}_4)_3}{342.17 \text{ g Al}_2(\text{SO}_4)_3}\right) = 0.275 \text{ mol Al}_2(\text{SO}_4)_3$

b)

$(37.5 \text{ g Al}_2(\text{SO}_4)_3)\left(\dfrac{1 \text{ mol Al}_2(\text{SO}_4)_3}{347.12 \text{ g Al}_2(\text{SO}_4)_3}\right)\left(\dfrac{12 \text{ mol O}}{1 \text{ mol Al}_2(\text{SO}_4)_3}\right)\left(\dfrac{6.022 \times 10^{23} \text{ O atoms}}{1 \text{ mol O}}\right) = 7.92 \times 10^{23} \text{ O atoms}$

Module 4
The Mole Concept

Introduction

In this module we will examine several equations that are used in problems involving the mole concept. The goal of this module is to teach you:

  1. how to interpret, use, and perform all of the important calculations that involve the mole

You will need a periodic table as you work all of these exercises. Atomic weights come directly from the periodic table.

Module 4 Key Equations & Concepts

1. **Molar mass $= \sum$ atomic weights of atoms in a compound, molecule, or ion**

   The molar mass, molecular weight, or formula weight[*] is calculated by summing the atomic weights of the atoms in the compound. This value gives the mass in grams of one mole of a substance.

2. **One mole $= 6.022 \times 10^{23}$ particles**

   Avogadro's relationship is used to convert from the number of moles of a substance to the number of atoms, ions, or molecules of that substance and vice versa.

3. **mass of one atom of an element** $= \left(\dfrac{\text{grams of an element}}{\text{1 mole of an element}} \right) \left(\dfrac{\text{1 mole of atoms}}{6.022 \times 10^{23} \text{ atoms}} \right)$

   The mass of one atom, ion, or molecule is used to determine the mass of a few atoms, ions, or molecules of a substance. Notice that the fraction in the first set of parentheses simply represents molar mass

4. **mole ratio**

   The chemical formula of a compound indicates the *ratio* of the different types of atom in the compound. The mole ratio can be used to convert from mass or moles of a compound to mass or moles of a specific atom in the compound.

---

[*]The terms molar mass, molecular weight, and formula weight all apply to the same concept/calculation. Technically, the term molecular weight should be used only with covalent compounds and formula weight applies only to ionic compounds. The more generic term *molar mass* is used frequently in chemical literature.

Sample Exercises
Determining the Molar Mass
1. What is the molar mass (formula weight) of calcium phosphate, Ca₃(PO₄)₂?
   The correct answer is: 310.2 g/mol

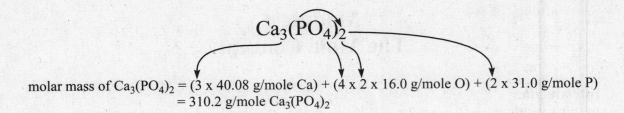

molar mass of $Ca_3(PO_4)_2$ = (3 x 40.08 g/mole Ca) + (4 x 2 x 16.0 g/mole O) + (2 x 31.0 g/mole P)

= 310.2 g/mole $Ca_3(PO_4)_2$

Determining the Number of Moles

2. *How many moles of calcium phosphate are there in 65.3 g of $Ca_3(PO_4)_2$?*
   The correct answer is: 0.211 mol $Ca_3(PO_4)_2$

$$? \text{ moles of } Ca_3(PO_4)_2 = 65.3 \text{ g } Ca_3(PO_4)_2 \left(\frac{1 \text{ mol } Ca_3(PO_4)_2}{310.2 \text{ g } Ca_3(PO_4)_2} \right)$$

$$= 0.211 \text{ mol } Ca_3(PO_4)_2$$

Molar mass of calcium phosphate from exercise #1.

Once the number of moles of the sample is known, we can determine the number of molecules or formula units of the substance. (Molecules are found in covalent compounds. Ionic compounds do not have molecules thus their smallest subunits are named formula units.)

Determining the Number of Molecules or Formula Units

3. *How many formula units of calcium phosphate are there in 0.211 moles of $Ca_3(PO_4)_2$?*

$$? \text{ formula units of } Ca_3(PO_4)_2 = 0.211 \text{ moles of } Ca_3(PO_4)_2 \left(\frac{6.022 \times 10^{23} \text{ formula units}}{1 \text{ mole of } Ca_3(PO_4)_2} \right)$$

$$= 1.27 \times 10^{23} \text{ formula units of } Ca_3(PO_4)_2$$

Avagadro's relationship

Be careful with the labels! You have just calculated the number of *formula units*. Do not confuse this with the number of atoms or the number of ions! All are valid questions with different answers!

Determining the Number of Atoms or Ions
4. *How many oxygen, O, atoms are there in 0.211 moles of Ca₃(PO₄)₂?*
   The correct answer is: 1.02 x 10²⁴ oxygen atoms

Using the last idea from the key concepts box, we can determine the mass of a few molecules or formula units of a compound.

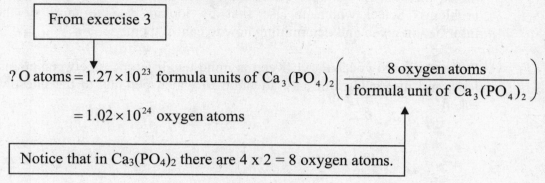

From exercise 3

$$? \text{ O atoms} = 1.27 \times 10^{23} \text{ formula units of } Ca_3(PO_4)_2 \left(\frac{8 \text{ oxygen atoms}}{1 \text{ formula unit of } Ca_3(PO_4)_2} \right)$$

$$= 1.02 \times 10^{24} \text{ oxygen atoms}$$

Notice that in $Ca_3(PO_4)_2$ there are 4 x 2 = 8 oxygen atoms.

Determining the Mass Molecules or Formula Units of a Substance
5. *What is the mass, in grams, of 25.0 formula units of Ca₃(PO₄)₂?*
   The correct answer is: 1.29 x 10⁻²⁰ g

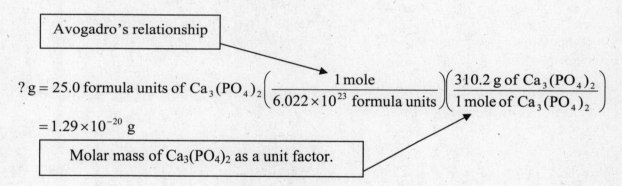

Avogadro's relationship

$$? \text{ g} = 25.0 \text{ formula units of } Ca_3(PO_4)_2 \left(\frac{1 \text{ mole}}{6.022 \times 10^{23} \text{ formula units}} \right) \left(\frac{310.2 \text{ g of } Ca_3(PO_4)_2}{1 \text{ mole of } Ca_3(PO_4)_2} \right)$$

$$= 1.29 \times 10^{-20} \text{ g}$$

Molar mass of $Ca_3(PO_4)_2$ as a unit factor.

Combined Equations
6. *How many carbon, C, atoms are there in 0.375 g of C₄H₈O₂?*
   The correct answer is: 1.03x 10²² atoms

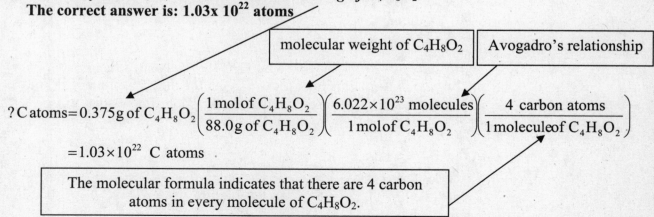

molecular weight of $C_4H_8O_2$ Avogadro's relationship

$$? \text{ C atoms} = 0.375 \text{ g of } C_4H_8O_2 \left(\frac{1 \text{ mol of } C_4H_8O_2}{88.0 \text{ g of } C_4H_8O_2} \right) \left(\frac{6.022 \times 10^{23} \text{ molecules}}{1 \text{ mol of } C_4H_8O_2} \right) \left(\frac{4 \text{ carbon atoms}}{1 \text{ molecule of } C_4H_8O_2} \right)$$

$$= 1.03 \times 10^{22} \text{ C atoms}$$

The molecular formula indicates that there are 4 carbon atoms in every molecule of $C_4H_8O_2$.

One of the problems most commonly encountered by students is figuring out where to start when working these problems. If you have trouble getting started, focus on the information that was given. All of the examples in this module began by using the mass or number of moles stated in the question. You will almost always use some combination of molar masses, Avogadro's relationship, and mole ratio to solve these problems. Select which to use first by looking at the units of the information given and determining how to cancel them.

Pay attention to vocabulary! Keep in mind the differences between atoms, ions, and molecules, and pay attention to which pertains to the question asked.

Module 5 Predictor Questions

The following questions may help you to determine to what extent you need to study this module. The questions are ranked according to ability.

Level 1 = basic proficiency
Level 2 = mid level proficiency
Level 3 = high proficiency

If you can correctly answer the Level 3 questions, then you probably do not need to spend much time with this module. If you are only able to answer the Level 1 problems, then you should review the topics covered in this module.

Level 1   1. Balance the following equations with the **smallest whole number coefficients**.

a) ___$Fe(NO_3)_3$ + ___NH_3 + ___H_2O → ___$Fe(OH)_3$ + ___NH_4NO_3

b) ___$C_2H_8N_2$ + ___N_2O_4 → ___N_2 + ___CO_2 + ___H_2O

Level 1   2. Given the balanced chemical reaction: $SiCl_4$ + 2Mg → Si + 2$MgCl_2$, how many grams of Si could be produced by reacting 1.46 kg of $SiCl_4$ with excess Mg?

Level 1   3. If 58 moles of NH_3 are combined with 32 moles of sulfuric acid, what is the limiting reactant and how much of the excess reactant is left over?

$$2NH_3 + H_2SO_4 → (NH_4)_2SO_4$$

Level 1   4. What is the percent yield if 28.50 g of FeO reacts with excess CO and produces 17.841 g of Fe?

$$FeO + CO → Fe + CO_2$$

Level 1   5. What volume of 0.158 M HBr solution is required to react completely with 38.77 mL of 0.226 M $Ca(OH)_2$ in the following reaction:

$$Ca(OH)_2 + 2HBr → CaBr_2 + 2 H_2O$$

Level 2   6. How many mL of 5.44 M $Sr(OH)_2$ are required to make 100.99 mL of a 0.189 M $Sr(OH)_2$ solution? What is the molar concentration of the Sr^{2+} ions in the 0.189 M solution? What is the molar concentration of the OH^- ions in the 0.189 M solution?

Module 5 Predictor Question Solutions

1. $1\ Fe(NO_3)_3 + 3\ NH_3 + 3\ H_2O \rightarrow 1\ Fe(OH)_3 + 3\ NH_4NO_3$

    $1\ C_2H_8N_2 + 2\ N_2O_4 \rightarrow 3\ N_2 + 2\ CO_2 + 4\ H_2O$

2.
$$(1.46\ kg\ SiCl_4)\left(\frac{10^3\ g\ SiCl_4}{1\ kg\ SiCl_4}\right)\left(\frac{1\ mol\ SiCl_4}{169.89\ g\ SiCl_4}\right)\left(\frac{1\ mol\ Si}{1\ mol\ SiCl_4}\right)\left(\frac{28.09\ g\ Si}{1\ mol\ Si}\right) = 2.41 \times 10^2\ g\ Si$$

3.
$$(58\ mol\ NH_3)\left(\frac{1\ mol\ (NH_4)_2SO_4}{2\ mol\ NH_3}\right) = 29\ mol\ (NH_4)_2SO_4$$

$$(32\ mol\ H_2SO_4)\left(\frac{1\ mol\ (NH_4)_2SO_4}{1\ mol\ H_2SO_4}\right) = 32\ mol\ (NH_4)_2SO_4$$

NH_3 is the limiting reagent. **29 mol $(NH_4)_2SO_4$ are made**.

$$(58\ mol\ NH_3)\left(\frac{1\ mol\ H_2SO_4}{2\ mol\ NH_3}\right) = 29\ mol\ H_2SO_4\ used$$

$32\ mol - 29\ mol = 3\ mol\ H_2SO_4$ remaining

4.
$$(28.50\ g\ FeO)\left(\frac{1\ mol\ FeO}{71.85\ g\ FeO}\right)\left(\frac{1\ mol\ Fe}{1\ mol\ FeO}\right)\left(\frac{55.85\ g\ Fe}{1\ mol\ Fe}\right) = 22.15\ g\ Fe = \text{theoretical yield}$$

$$\% \text{ yield} = \left(\frac{\text{actual yield}}{\text{theoretical yield}}\right)100 = \left(\frac{17.841\ g\ FeO}{22.15\ G\ FeO}\right)100 = 80.55\%$$

5.
$$(38.77\ mL)\left(\frac{10^{-3}\ L}{1\ mL}\right)\left(\frac{0.226\ mol\ Ca(OH)_2}{1\ L}\right)\left(\frac{2\ mol\ HBr}{1\ mol\ Ca(OH)_2}\right)\left(\frac{1\ L}{0.158\ mol\ HBr}\right)\left(\frac{1\ mL}{10^{-3}\ L}\right) = 0.111\ mL$$

6.

$$M_1V_1 = M_2V_2$$

$$(5.44M)(V_1) = (0.189\,M)(100.99mL)$$

$$V_1 = 3.51\,mL$$

$$\left(\frac{0.180\,mol\,Sr(OH)_2}{1\,L}\right)\left(\frac{1\,mol\,Sr^{2+}}{1\,mol\,Sr(OH)_2}\right) = 0.189\,M\,Sr^{2+}$$

$$\left(\frac{0.180\,mol\,Sr(OH)_2}{1\,L}\right)\left(\frac{2\,mol\,OH^-}{1\,mol\,Sr(OH)_2}\right) = 0.378\,M\,OH^-$$

Module 5
Chemical Reaction Stoichiometry

Introduction
In this module we will look at several problems that involve reaction stoichiometry. The important points to learn in this module are:

1. balancing chemical reactions
2. basic reaction stoichiometry
3. limiting reactant calculations
4. percent yield calculations
5. reactions in solution

You will need a periodic table to calculate molecular weights in these problems.

Module 5 Key Equations & Concepts

1. **Percent yield**

$$\% \text{ yield} = \frac{\text{actual yield}}{\text{theoretical yield}} \times 100$$

The percent yield formula is used to determine the percentage of the theoretical yield that was formed in a reaction.

2. *Molarity (M)*

M = **moles solute/L solution**

M **x L = moles or** *M* **x mL = mmol**

The relationship of molarity and volume is used to convert from solution concentrations to moles or from volume of a solution to moles of a solution.

Chemical reactions symbolize what happens when chemical substances are mixed and new substances are formed. Before proceeding, it is important to review some vocabulary:

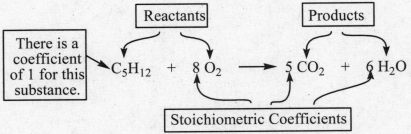

Reactants are the chemical species with which the reaction begins. Products are the species resulting from the reaction. Stoichiometric coefficients are required to "balance" the equation. This process insures that equal numbers of atoms of each element are present on both sides of the reaction. Otherwise the reaction would violate the Law of Conservation of Mass.

There are numerous ways to approach balancing an equation. It does not matter where you start. However, it is often easiest to use the following steps:

1. If possible, start with an element that appears in only one compound on each side of the equation.
2. Save balancing anything that appears without other elements (O_2, Fe(s), etc.) for last.
3. If the equation contains polyatomic ions, you may try looking at them as whole entities and balancing them as such rather than looking at each individual type of atom.

Sample Exercises
Balancing Chemical Reactions

1. **Balance this chemical reaction using the smallest whole numbers.**

$$Ca(OH)_2 + H_3PO_4 \rightarrow Ca_3(PO_4)_2 + H_2O$$

Consider starting with Ca since it appears in only one compound on each side of the reaction. It is probably easiest to balance (PO_4^{3-}) as the polyatomic ion rather than individually as P and O atoms. Then, only H and O are left.

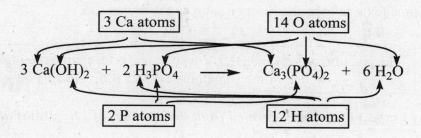

2. **Balance this chemical reaction using the smallest whole numbers.**

$$C_6H_{14} + O_2 \rightarrow CO_2 + H_2O$$

Start with C, and save oxygen to balance last!

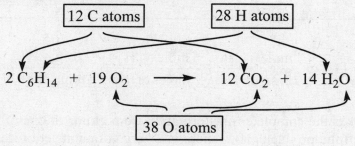

Simple Reaction Stoichiometry
Properly balanced chemical equations are very important because they allow calculations pertaining to chemical reactions. This is called reaction stoichiometry.

3. How many moles of H_2 can be formed from the reaction of 3.0 moles of Na with excess H_2O.
  The correct answer is: 1.5 moles H_2

$$2\,Na \;+\; 2\,H_2O \;\rightarrow\; 2\,NaOH \;+\; H_2$$

INSIGHT:
The word **excess** is important. It is your clue that this problem does **not** involve a limiting reactant calculation.
The reaction ratio, which comes from the balanced reaction, is 2 moles of Na consumed for every 1 mole of H_2 formed. Write it as a unit factor.

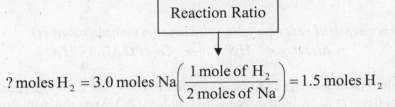

$$? \text{ moles } H_2 = 3.0 \text{ moles Na}\left(\frac{1\,\text{mole of } H_2}{2\,\text{moles of Na}}\right) = 1.5 \text{ moles } H_2$$

The reaction ratio is a new conversion factor that relates moles of any reactant or product to moles of another reactant or product. The reaction ratio comes from the balanced equation. (Some texts refer to the reaction ratio as the mole ratio.)

INSIGHT:
The reaction ratio is the ONLY way to use information about one species in the reaction to determine something about a *different species* in the reaction.

4. How many grams of H_2 can be formed from the reaction of 11.2 grams of Na with excess H_2O?
  The correct answer is: 0.494 g H_2

$$2\,Na \;+\; 2\,H_2O \;\rightarrow\; 2\,NaOH \;+\; H_2$$

Converts g of Na to moles of Na	Reaction Ratio	Converts moles of H_2 to g of H_2

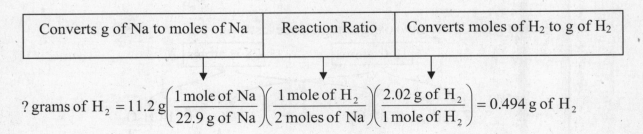

$$? \text{ grams of } H_2 = 11.2 \text{ g}\left(\frac{1\,\text{mole of Na}}{22.9\,\text{g of Na}}\right)\left(\frac{1\,\text{mole of } H_2}{2\,\text{moles of Na}}\right)\left(\frac{2.02\,\text{g of } H_2}{1\,\text{mole of } H_2}\right) = 0.494 \text{ g of } H_2$$

This problem makes the complete transformation from grams of one of the reactants, Na, to grams of one of the products, H_2. There is a very common set of transformations that are used in this calculation which will be used in many reaction stoichiometry problems.

  grams of X → moles of X → reaction ratio → moles of Y → grams of Y

Notice how all of the units cancel, leaving g of H_2.

Limiting Reactants

5. *What is the maximum number of grams of H_2 that can be formed from the reaction of 11.2 grams of Na with 9.00 grams of H_2O?*
   The correct answer is: 0.494 g of H_2

$$2\ Na + 2\ H_2O \rightarrow 2\ NaOH + H_2$$

INSIGHT: | The word excess is not in this problem, and amounts of both reactants are given. These are your *clues that this is a limiting reactant problem.*

You must perform reaction stoichiometry steps for every reactant for which an amount was given in the problem. In this case that is two steps.

$$? \text{grams of } H_2 = 11.2\,\text{g of Na}\left(\frac{1\,\text{mole of Na}}{22.9\,\text{g of Na}}\right)\left(\frac{1\,\text{mole of } H_2}{2\,\text{moles of Na}}\right)\left(\frac{2.02\,\text{g of } H_2}{1\,\text{mole of } H_2}\right) = 0.494\,\text{g of } H_2$$

$$? \text{grams of } H_2 = 9.00\,\text{g of } H_2O\left(\frac{1\,\text{mole of } H_2O}{18.0\,\text{g of } H_2O}\right)\left(\frac{1\,\text{mole of } H_2}{2\,\text{moles of } H_2O}\right)\left(\frac{2.02\,\text{g of } H_2}{1\,\text{mole of } H_2}\right) = 0.505\,\text{g of } H_2$$

 YIELD | The maximum amount will be the _smallest_ amount that you calculate in the reaction stoichiometry steps!

This calculation indicates that all 11.2 g of Na are used in the production of 0.494 g of H_2. Since there is no Na left, no more H_2 can be produced, even though there is still H_2O remaining. Once one reactant is completely used, no more product can be made. In this example, Na is the *limiting reactant* and H_2O is the *excess reactant.*

Percent Yield

6. *If 11.2 g of Na reacts with 9.00 g of H_2O and 0.400 g of H_2 is formed, what is the percent yield of the reaction?*
   The correct answer is: 81.0%

> This is the actual yield.

$$2\ Na + 2\ H_2O \rightarrow 2\ NaOH + H_2$$

INSIGHT: | Key clues that indicate percent yield problems are: a) amounts of both reactants given, b) an amount for the product, and c) the words percent yield.

In percent yield problems, limiting reactant calculations are frequently performed first to determine the *theoretical yield.* This is the amount of product that is formed if the reaction goes 100% to completion (this rarely happens in the lab!) and what was calculated in exercises 4 and 5. For this problem the theoretical yield is the same as determined in exercise 5 (0.494 g of H_2).

$$\% \text{ yield} = \frac{\text{actual yield}}{\text{theoretical yield}} \times 100 = \frac{0.400 \text{ g}}{0.494 \text{ g}} \times 100\% = 81.0\%$$

Reactions in Solution

7. How many mL of 0.250 M HCl are required to react with 15.0 mL of 0.150 M Ba(OH)₂?

The correct answer is: 18.0 mL HCl

$$2 \text{ HCl}(aq) + Ba(OH)_2(aq) \rightarrow BaCl_2(aq) + 2 \text{ H}_2O(\ell)$$

INSIGHT: Key clues to indicate a reaction in solution are the presence of solution concentration(s) (0.250 M and 0.150 M) and volume(s) in the problem, in addition to a balanced chemical equation.

mL x M = millimoles Ba(OH)₂	Reaction Ratio	mmol HCl x (1/M) = mL HCl

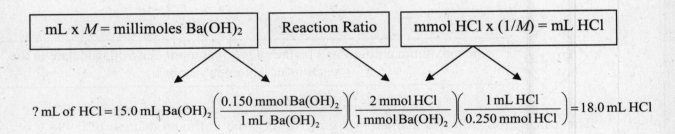

$$? \text{ mL of HCl} = 15.0 \text{ mL Ba(OH)}_2 \left(\frac{0.150 \text{ mmol Ba(OH)}_2}{1 \text{ mL Ba(OH)}_2} \right) \left(\frac{2 \text{ mmol HCl}}{1 \text{ mmol Ba(OH)}_2} \right) \left(\frac{1 \text{ mL HCl}}{0.250 \text{ mmol HCl}} \right) = 18.0 \text{ mL HCl}$$

Module 6 Predictor Questions

The following questions may help you to determine to what extent you need to study this module. The questions are ranked according to ability.

    Level 1 = basic proficiency
    Level 2 = mid level proficiency
    Level 3 = high proficiency

If you can correctly answer the Level 3 questions, then you probably do not need to spend much time with this module. If you are only able to answer the Level 1 problems, then you should review the topics covered in this module.

Level 1   1. Determine **all** of the reaction types that will correctly classify the following reaction.
$$2NH_4NO_3 \, (s) \rightarrow 2N_2 \, (g) + O_2 \, (g) + 4H_2O \, (g)$$

Level 2   2. Determine **all** of the reaction types that will correctly classify the following reaction.
$$AgNO_3 \, (aq) + HCl(aq) \rightarrow AgCl(s) + HNO_3(aq)$$

Level 3   3. Determine **all** of the reaction types that will correctly classify the following reaction.
$$BaCO_3 \rightarrow BaO + CO_2$$

Level 2   4. Predict the products of the following reactions:
a) $Cu(NO_3)_2 + Na_2S \rightarrow$???
b) $CdSO_4 + H_2S \rightarrow$???
c) $Ba(NO_3)_2 + K_2CO_3 \rightarrow$???

Level 1   5. What is the **total** ionic equation for the following formula unit equation?
$$BaCl_2 \, (aq) + Na_2SO_4 \, (aq) \rightarrow BaSO_4(s) + 2NaCl(aq)$$

Level 2   6. What is the **net** ionic equation for the reaction of H_3PO_4 with NaOH?

Level 2   7. Write the **total** and **net** ionic equations for the reaction of Zn solid with $AgNO_3$.

Module 6 Predictor Question Solutions

1. This is a decomposition reaction, an oxidation-reduction reaction, and a gas forming reaction.

2. This is both a metathesis and a precipitation reaction.

3. This is a decomposition reaction and a gas forming reaction.

4.   a) $Cu(NO_3)_2 + Na_2S \rightarrow$ **$NaNO_3$ (aq)** **+ CuS(s)**
       b) $CdSO_4 + H_2S \rightarrow$ **H_2SO_4 (aq) + CdS(s)**
       c) $Ba(NO_3)_2 + K_2CO_3 \rightarrow$ **KNO_3(aq) + $BaCO_3$(s)**

5. $BaCl_2$ (aq) + Na_2SO_4 (aq) $\rightarrow BaSO_4$ (s) + 2NaCl (aq)
   The *total* ionic equation is:
   Ba^{2+}(aq) + 2Cl$^-$(aq) + 2Na$^+$(aq) + $SO_4{}^{2-}$ (aq) $\rightarrow$ $BaSO_4$ (s) + 2Na$^+$(aq) + 2Cl$^-$(aq)

6. The complete molecular equation is:
   $$H_3PO_4 \text{ (aq)} + 3NaOH\text{(aq)} \rightarrow Na_3PO_4\text{(aq)} + 3H_2O(\ell)$$

   The *total* ionic equation is:
   H_3PO_4 (aq) + 3Na$^+$ (aq) + 3OH$^-$(aq) $\rightarrow$ 3Na$^+$ (aq) + $PO_4{}^{3-}$ (aq) + $3H_2O(\ell)$
   Note that the *weak* acid H_3PO_4 remains intact.

   The *net* ionic equation is:
   H_3PO_4 (aq) + 3OH$^-$(aq) $\rightarrow$ $PO_4{}^{3-}$ (aq) + $3H_2O(\ell)$
   (The spectator ions, Na$^+$, were removed.)

7. The complete molecular equation is: $Zn(s) + 2AgNO_3$ (aq) $\rightarrow 2Ag(s) + Zn(NO_3)_2$ (aq)

   The *total* ionic equation is:
   $Zn(s) + 2Ag^+$ (aq) + 2NO$_3^-$ (aq) $\rightarrow 2Ag(s) + Zn^{2+}$ (aq) + 2NO$_3^-$ (aq)

   The *net* ionic equation is:
   $Zn(s) + 2Ag^+$ (aq) $\rightarrow 2Ag(s) + Zn^{2+}$ (aq)

Module 6
Types of Chemical Reactions

Introduction

This module focuses on recognizing several types of chemical reactions and predicting their products. The objectives of this module are to learn:

. how to use the reactants of a chemical reaction to discern the type of reaction
. to predict the reaction products of metathesis reactions
. how to write the total and net ionic equations for reactions

---

Module 6 Key Equations & Concepts

1. **Formula Unit Equations**

   Formula unit equations show all of the species involved in a reaction as ionic or molecular *compounds*: $KOH(aq) + HI(aq) \rightarrow KI(aq) + H_2O(\ell)$

2. **Total Ionic Equations**

   Total ionic equations show all of the ions in their ionized states in solution. All species that ionize completely in water are shown as separated ions:

   $K^+(aq) + OH^-(aq) + H^+(aq) + I^-(aq) \rightarrow K^+(aq) + I^-(aq) + H_2O(\ell)$

   Note that all gases, solids, and liquids are left intact.

3. **Net Ionic Equations**

   To write net ionic equations, remove all spectator ions from the total ionic equation. Spectator ions are species that do not change as the reaction proceeds from reactants to products: $OH^-(aq) + H^+(aq) \rightarrow H_2O(\ell)$

Sample Exercises

Reduction-Oxidation Reactions

Reduction-oxidation reactions are those in which electrons are transferred from one species to another. Reductions cannot occur without accompanying oxidations, so these are often called *redox reactions*. Your textbook has a series of rules for assigning oxidation numbers to elements in chemical species. If you do not know the rules for oxidation states, learn them now.

1. What reaction types are represented by this chemical reaction?

$$2\,Na(s)\ +\ 2\,H_2O(\ell) \rightarrow\ 2\,NaOH(aq)\ +\ H_2(g)$$

The correct answer is: This is a redox reaction (we will see later in this module that it is also a combination reaction).

41

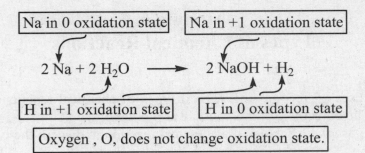

H: +1 oxidation state → 0 oxidation state = reduction
Na: 0 oxidation state → +1 oxidation state = oxidation

INSIGHT: To recognize redox reactions you must *look for chemical species that are changing their oxidation states*.

 Notice that Na in the above reaction is in its *elemental state* on the reactant side of the reaction and in a compound on the other side (the same is true of H). This is a big clue that you are dealing with a redox reaction. All species in their elemental states have oxidation states of zero, and species in compounds typically do not have oxidation states of zero. Thus, the oxidation state probably changes during the reaction!

Combination Reactions
2. What reaction types are represented by this chemical reaction?

$$3 \, Sr(s) \; + \; N_2(g) \; \rightarrow \; Sr_3N_2(s)$$

The correct answer is: This is both a combination reaction and a redox reaction.

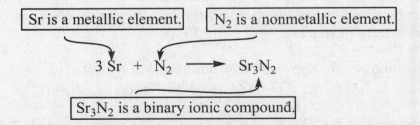

INSIGHT: Combination reactions are characterized by *a) the reaction of two elements to form a compound, b) the reaction of a compound and an element to form a new compound, or c) the reaction of two compounds to form a new compound.*

Combination reactions may also frequently be classified as another reaction type. In this case the second classification is a redox reaction.

Sr: 0 → +2 oxidation state = oxidation
N: 0 → -3 oxidation state = reduction

Decomposition Reactions

3. What reaction types are represented by this chemical reaction?

$$2\ CaO(s)\ \rightarrow\ 2\ Ca(s)\ +\ O_2(g)$$

The correct answer is: This is a decomposition reaction and a redox reaction.

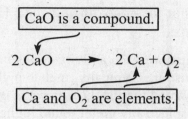

INSIGHT:

> There are three types of decomposition reactions: *a) compounds decomposing into two or more elements, b) compounds decomposing into another compound and an element, and c) compounds decomposing into two simpler compounds.*

Decomposition reactions are the reverse of combination reactions. Instead of putting elements or compounds together to make new compounds decomposition reactions break compounds into elements or less complex compounds.

As in combination reactions, decomposition reactions can frequently be classified as other reaction types, in this case a redox reaction.

Ca: +2 → 0 oxidation state = reduction
O: -2 → 0 oxidation state = oxidation

Displacement Reactions

3. What reaction types are represented by this reaction?

$$2\ Al(s)\ +\ 3\ H_2SO_4(aq)\ \rightarrow\ Al_2(SO_4)_3(aq)\ +\ 3\ H_2(g)$$

The correct answer is: This is a displacement reaction and a redox reaction.

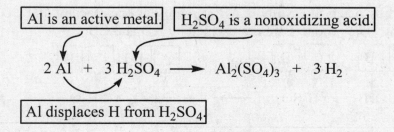

43

Displacement reactions are characterized by *one element replacing a second element in a compound.* The three types of displacement reactions are: *a) an active metal displacing the metal from a less active metal's salt, b) an active metal displacing hydrogen from either HCl or H_2SO_4, and c) an active nonmetal displacing the nonmetal from a less active nonmetal's salt.*

Displacement reactions involve the reaction of metals or nonmetals on the activity series with salts of less active metals or the nonoxidizing acids HCl and H_2SO_4. HNO_3 is the most common oxidizing acid.

TIP — If you are not familiar with the activity series in your text, be certain that you understand how to use it. Typically, metals higher up on the list can displace any metal lower down on the list. *The reverse is not true.* Metals found lower on the list *cannot* displace any metal found higher up on the activity series.

Metathesis Reactions

5. *What reaction types are represented by this reaction?*

$$Ba(OH)_2(aq) \;+\; H_2SO_4(aq) \;\rightarrow\; BaSO_4(s) \;+\; 2\,H_2O(\ell)$$

The correct answer is: This is a metathesis reaction that is both an acid-base and a precipitation reaction.

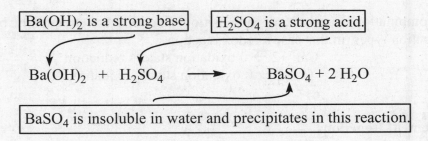

$Ba(OH)_2$ is a strong base. H_2SO_4 is a strong acid.

$$Ba(OH)_2 \;+\; H_2SO_4 \;\longrightarrow\; BaSO_4 + 2\,H_2O$$

$BaSO_4$ is insoluble in water and precipitates in this reaction.

Metathesis reactions are characterized by *the reactants switching their anions.*

This is exhibited by using the symbols AB to represent one reactant and CD to represent the other reactant. The products are represented by AD and CB.

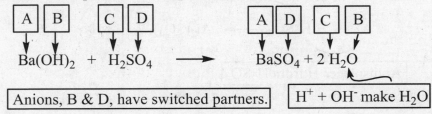

A B   C D   A D   C B
$$Ba(OH)_2 \;+\; H_2SO_4 \;\longrightarrow\; BaSO_4 + 2\,H_2O$$

Anions, B & D, have switched partners. $H^+ + OH^-$ make H_2O

When an acid reacts with a base, both a salt ($BaSO_4$ in this case) and water (if the base is a hydroxide) will be formed. Water is formed by the combination of the H^+ with the OH^- when the anions switch partners.

> **INSIGHT:** Precipitation reactions are characterized by *the formation of a compound that is insoluble in water*.

 You must understand and use the solubility rules from your textbook to recognize a precipitation reaction since the phases of the product compounds are not frequently given, as you will see in the exercises below.

Predicting Products of Metathesis Reactions

6. *What are the products of this chemical reaction?*

$$Sr(OH)_2(aq) \ + \ Fe(NO_3)_3(aq) \rightarrow ??? + ???$$

The correct answer is: $Sr(NO_3)_2$ and $Fe(OH)_3$

$\boxed{Sr^{2+} \text{ and } OH^-}$ $\boxed{Fe^{3+} \text{ and } NO_3^-}$

$3 \ Sr(OH)_2 \ + \ 2 \ Fe(NO_3)_3 \longrightarrow 3 \ Sr(NO_3)_2 \ + \ 2 \ Fe(OH)_3$

$\boxed{\text{Notice that the reaction is balanced.}}$ $\boxed{Sr^{2+} \text{ and } NO_3^-}$ $\boxed{Fe^{3+} \text{ and } OH^-}$

> **INSIGHT:** The anions have switched partners forming new chemical compounds. ***Basic rules of ionic compound formation must be obeyed. Thus, the total charge of the positive ions is equal to the total charge of the negative ions, resulting in the formation of neutral compounds.***

Total and Net Ionic Equations

Net ionic equations are very helpful because they allow you to focus on the essential parts of the reaction. For example, net ionic equations can make assigning oxidation numbers much easier.

7. *Write the total ionic and net ionic equations for this reaction.*

$$Ba(OH)_2(aq) \ + \ 2 \ HCl(aq) \ \rightarrow \ BaCl_2(aq) \ + \ 2 \ H_2O(\ell)$$

The correct total ionic equation is:

$$Ba^{2+}(aq) + 2 \ OH^-(aq) + 2 \ H^+(aq) + 2 \ Cl^-(aq) \rightarrow Ba^{2+}(aq) + 2 \ Cl^-(aq) + 2 \ H_2O(\ell)$$

The correct net ionic equation is:

$$2 \ OH^-(aq) + 2 \ H^+(aq) \rightarrow 2 \ H_2O(\ell)$$

or

$$OH^-(aq) + H^+(aq) \rightarrow H_2O(\ell)$$

These problems are very difficult if you are not readily familiar with the solubility rules.

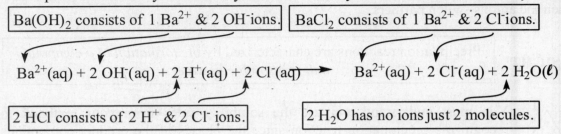

Ba(OH)$_2$ consists of 1 Ba^{2+} & 2 OH$^-$ions. | BaCl$_2$ consists of 1 Ba^{2+} & 2 Cl$^-$ions.

Ba^{2+}(aq) + 2 OH$^-$(aq) + 2 H$^+$(aq) + 2 Cl$^-$(aq) $\longrightarrow$ Ba^{2+}(aq) + 2 Cl$^-$(aq) + 2 H$_2$O(ℓ)

2 HCl consists of 2 H$^+$ & 2 Cl$^-$ ions. | 2 H$_2$O has no ions just 2 molecules.

The 2 in front of the OH$^-$ comes from the subscript 2 in Ba(OH)$_2$. The 2's in front of H$^+$ and Cl$^-$ come from the stoichiometric coefficient 2.

TIP — If there are both a subscript and a coefficient, multiply them together to determine the number of ions present.

INSIGHT: *Spectator ions do not change from reactant to product.* Ba^{2+} and Cl$^-$ are spectator ions in this reaction. Once the correct total ionic equation is written, *removal of the spectator ions, Ba^{2+} and Cl$^-$, leaves the correct net ionic equation.*

$$2\ OH^-(aq) + 2\ H^+(aq) \rightarrow 2\ H_2O(\ell)$$
or
$$OH^-(aq) + H^+(aq) \rightarrow H_2O(\ell)$$

INSIGHT: Just as discussed in Module 5, we must reduce the stoichiometric coefficients to their smallest whole numbers.

8. Write the total and net ionic equations for this reaction.
$$NaOH(aq) + CH_3COOH(aq) \rightarrow NaCH_3COO(aq) + H_2O(\ell)$$

The correct total ionic equation is:
$$Na^+(aq) + OH^-(aq) + CH_3COOH(aq) \rightarrow Na^+(aq) + CH_3COO^-(aq) + H_2O(\ell)$$

The correct net ionic equation is:
$$OH^-(aq) + CH_3COOH(aq) \rightarrow CH_3COO^-(aq) + H_2O(\ell)$$

NaOH is a **strong** water soluble base that ionizes into Na^+ and OH^- ions in aqueous solutions.	NaCH₃COO is a **water soluble salt** that ionizes into Na^+ and CH_3COO^- ions in aqueous solutions.

$$Na(aq) + OH^-(aq) + CH_3COOH(aq) \longrightarrow Na^+(aq) + CH_3COO^-(aq) + H_2O(\ell)$$

CH₃COOH is a **weak** water soluble acid that ionizes so slightly in aqueous solutions that it is not separated into ions.	H_2O is a molecule and does not form ions.
The only spectator ion in this reaction is Na^+.	Removing the Na^+ ion from the total ionic equation leaves the net ionic equation.

$$OH^-(aq) + CH_3COOH(aq) \rightarrow CH_3COO^-(aq) + H_2O(\ell)$$

Notice that the net ionic equation tells us that a strong base, hydroxide ion, reacts with a weak acid, acetic acid, to form the acetate ion and water.

 YIELD | In total and net ionic equations there are three classes of chemical species that are broken into ions: *a) strong acids, b) strong bases, and c) water soluble salts.* Never break gases, liquids, or solids into ions. |

INSIGHT: | Total and net ionic equations are exceedingly difficult to write unless you are familiar with *a) the strong acids, b) the strong bases, c) the solubility rules, and d) how ionic compounds ionize in aqueous solutions.* |

Practice Test Two
Modules 4-6

Level 1   1. What is the mass of the oxygen atoms in 42.7 g of CH_3COOH?

Level 1   2. Balance the following reaction using the smallest possible whole number coefficients.

$$\_\_\_P_4O_{10} + \_\_\_H_2O \rightarrow \_\_\_H_3PO_4$$

Levels 1-2   3. A 28.42 g sample of silver nitrate is reacted with 14.00 g of calcium chloride. If 10.72 g of calcium nitrate is produced, then what is the percent yield for the reaction?

Level 3   4. Hydrochloric acid, HCl, is formed by the following sequential reactions. How many mols of HCl are formed from 105 g of H_2O if the % yields of steps one and two are 67.2% and 86.9%, respectively?

$$2H_2O \rightarrow 2H_2 + O_2 \quad (67.2\%)$$
$$H_2 + Cl_2 \rightarrow 2HCl \quad (86.9\%)$$

Level 2   5. In a given solution of $Fe(NO_3)_3$, the molar concentration of Fe^{3+} is 0.150 M and the molar concentration of NO_3^- is 0.450 M. To what volume must 250.0 mL of the $Fe(NO_3)_3$ solution be diluted to create a $Fe(NO_3)_3$ solution with a concentration of 0.0850 M?

Level 1   6. Excess $AlCl_3$ reacts with 52.3 mL of 0.500 M $AgNO_3$ according to the following reaction.

$$AlCl_3 + 3AgNO_3 \rightarrow 3AgCl + Al(NO_3)_3$$

What is the final volume if the final concentration of $Al(NO_3)_3$ is 0.0673 M?

Level 1   7. Determine *all* of the reaction types that classify the reactions given in questions 4 and 6.

Level 1   8. Write both the *total* and the *net* ionic equations for the reaction of chloric acid with stronium hydroxide.

Module 7 Predictor Questions

The following questions may help you to determine to what extent you need to study this module. The questions are ranked according to ability.

Level 1 = basic proficiency

Level 2 = mid level proficiency

Level 3 = high proficiency

If you can correctly answer the Level 3 questions, then you probably do not need to spend much time with this module. If you are only able to answer the Level 1 problems, then you should review the topics covered in this module.

Level 1   1. What is the electron configuration of tellurium, $_{52}$Te?

Level 2   2. What is the principal quantum number for the **valence** electrons of each element?
        a) K
        b) P
        c) Mn

Level 2   3. Choose the set of quantum numbers which would NOT be correct for any of the electrons in the ground state configuration of the element Si.
        a) $n = 3$, $\ell = 2$, $m_\ell = -1$, $m_s = +1/2$
        b) $n = 2$, $\ell = 1$, $m_\ell = -1$, $m_s = -1/2$
        c) $n = 3$, $\ell = 0$, $m_\ell = 0$, $m_s = +1/2$
        d) $n = 2$, $\ell = 0$, $m_\ell = 0$, $m_s = -1/2$
        e) $n = 1$, $\ell = 0$, $m_\ell = 0$, $m_s = -1/2$

Level 3   4. Answer the following questions regarding Fe
        a) How many d electrons are there in the ground state electron configuration?
        b) What is the value of the **n** quantum number for the d electrons in Fe?
        c) How many of the d electrons in Fe are *paired*?
        d) How many of the d electrons in Fe are *unpaired*?

Level 3   5. What is the maximum number of electrons in an atom that could be described by the following quantum number?
        a) $n = 3$, $\ell = 2$
        b) $n = 3$, $\ell = 1$
        c) $n = 5$, $\ell = 2$
        d) $n = 4$, $\ell = 3$
        e) $n = 2$, $\ell = 0$

Level 3   6. The orientation in space of an orbital is designated by which of the four
                quantum numbers?

Level 3   7. What is the value of the angular momentum quantum number for each
                of the following types of orbitals?
                a) s
                b) d
                c) p
                d) f

Module 7 Predictor Question Solutions

1. Te: $1s^2 2s^2 2p^6 3s^2 3p^6 4s^2 3d^{10} 4p^6 5s^2 4d^{10} 5p^4$ OR $[Kr]\ 5s^2 4d^{10} 5p^4$

2. a) K; $n = 4$
    b) P; $n = 3$
    c) Mn; $n = 3$

3. Statement a) is untrue. The combination of quantum numbers $n = 3$ and $\ell = 2$ indicates an electron located in the d block of the fourth row of the periodic table. Since Si is located in the p block of the third row, this electron cannot be found in Si.

4. The electron configuration for Fe is: $[Ar]4s^2 3d^6$
    a) There are six d electrons.
    b) $n = 3$
    c) and d) If Hund's rule is obeyed, then the d electrons must be placed into the five d orbitals such that each orbital contains one electron before any orbital can contain two electrons. Thus, there are **two paired electrons** and **four unpaired electrons.**

5. This question can be answered by looking only at the value of ℓ since it describes the type of orbital (s, p, d, or f). Remember that each orbital can hold up to two electrons.

    $\ell = 0$ indicates s orbitals (one per set)
    $\ell = 1$ indicates p orbitals (three per set)
    $\ell = 2$ indicates d orbitals (five per set)
    $\ell = 3$ indicates f orbitals (seven per set)

    a) $n = 3, \ell = 2 \rightarrow$ 10 electrons
    b) $n = 3, \ell = 1 \rightarrow$ 6 electrons
    c) $n = 5, \ell = 2 \rightarrow$ 10 electrons
    d) $n = 4, \ell = 3 \rightarrow$ 14 electrons
    e) $n = 2, \ell = 0 \rightarrow$ 2 electrons

6. The orientation of the orbital is given by m_ℓ.

7. As described in the solution to question 5, each type of orbital is described by a certain value of ℓ.

    a) $\ell = 0$
    b) $\ell = 2$
    c) $\ell = 1$
    d) $\ell = 3$

Module 7
Electronic Structure of Atoms

Introduction

This module describes the meaning of quantum numbers and how to assign them to electrons. The goals of this module are to explain:

1. how to determine the quantum numbers for an element
2. how to discern the correct atomic electronic structure of an element by looking at the periodic table
3. how to write the entire set of quantum numbers for an element

You will need access to a periodic table in order to do the sample exercises in this module.

Module 7 Key Equations & Concepts

1. **Principal quantum number**

   Represented by the symbol n, this quantum number describes the main energy level of an atom.

   $n = 1, 2, 3, 4, 5, 6, ...\infty$

2. **Angular momentum quantum number**

   Represented by the symbol ℓ, the angular momentum quantum number describes the shape of the atomic orbitals and the region of space occupied by electrons. The allowed values of ℓ are dependent on the value of n. Each value of ℓ corresponds to a specific type of orbital.

   $\ell = 0, 1, 2, 3, 4, ... (n-1)$
   $\ell = s, p, d, f, g, ... (n-1)$

3. **Magnetic quantum number**

   Represented by the symbol m_ℓ, the magnetic quantum number describes the number of atomic orbitals that are possible for each value of ℓ.

   $m_\ell = -\ell, -\ell+1, -\ell+2, ..., 0, ... \ell-2, \ell-1, \ell$

4. **Spin quantum number**

   Represented by the symbol m_s, the spin quantum number describes the relative magnetic orientation of the electrons in an atom. It also defines the maximum number of electrons that can occupy one orbital.

   $m_s = +1/2$ or $-1/2$

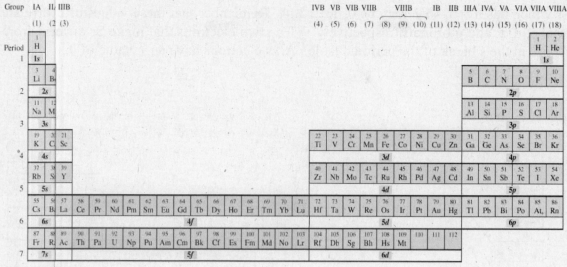

© 2004 Thomson/Brooks Cole

Sample Exercises

Principal Quantum Number

1. What is the value of the principal quantum number, n, for the valence electrons in a Sr atom?

The correct answer is: n = 5

INSIGHT:	Sr is on the 5th row of the periodic chart. All of the main group elements on the 5th row, except the transition metals, will have n = 5. The value of n is always equal to the period number except for transition metals, lanthanides, and actinides.

2. What is the value of the n quantum number for the valence electrons in a Zr atom?

The correct answer is: n = 4

INSIGHT:	Zr is a transition metal on the 5th row of the periodic chart. ***Transition metals have an n value that is 1 number less than the row where they appear on the periodic chart.*** Thus a 5th row transition metal has n = 4.

Orbital Angular Momentum Quantum Number

3. What is the value of the orbital angular momentum quantum number, ℓ, for the valence electrons in a Sr atom?

The correct answer is: ℓ = 0

Sr which has n = 5, so ℓ may be 0, 1, 2, 3, 4. Remember that these values of ℓcorrelate to s, p, d, f, and g orbitals, respectively.   The two electrons that make Sr different from Ar are in the s block of the periodic table. All s electrons have an ℓ value of 0.

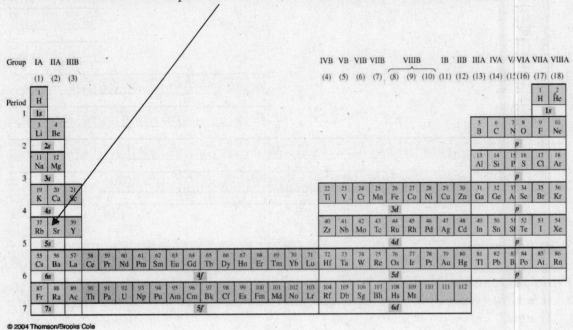

© 2004 Thomson/Brooks Cole

Each box on the periodic table represents one electron. You canuse this along with the knowledge that each orbital holds two electrons to elp you remember which part of the periodic table represents each type of cbital.

- The two electrons represented by the alkali and alkaline earth metal on each row of the periodic table represent one s orbital in each period.
- The six electrons represented by the elements boronthrough neon in row two represent three p orbitals. The same istrue for each period below period two.
- The ten electrons represented by the ten elements located in each row of transition metals represent five d orbitals.
- The fourteen elements in the lanthanide period and the ourteen in the actinide period represent electrons in seven f orbitls.

4. **What is the value of the orbital angular momentum quantum number, ℓ, fc the electrons which make a Zr atom different from a Sr atom?**
   The correct answer is: ℓ = 2 = d electrons

For Zn which has n = 5, ℓ may be 0, 1, 2, 3, or 4 (correlating to s, p, d, f, or g oritals).

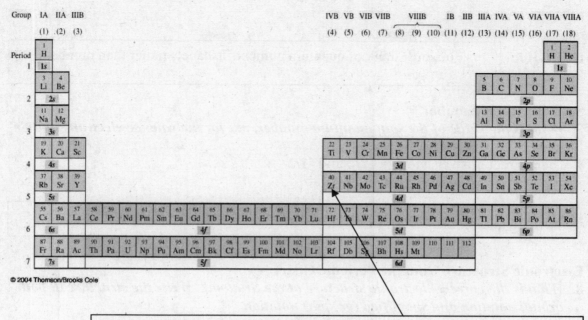

INSIGHT: The electrons which make a Zr atom, element 45, different from a Sr atom, element 38, are in the d block of the periodic table. *All d electrons have an ℓ value of 2.*

Magnetic Quantum Number

5. *What is the value of the magnetic quantum number, m_ℓ, for the valence electrons in a Sr atom?*
   The correct answer is: $m_\ell = 0$

For Sr, $\ell = 0$.

INSIGHT: For s electrons $\ell = 0$. The magnetic quantum number may have any integer value from $-\ell$ all the way to ℓ. If $\ell = 0$, then the only possible value of m_ℓ is 0.

6. *What is the value of the magnetic quantum number, m_ℓ, for the electrons which make a Zr atom different from a Sr atom?*
   The correct answer is: $m_\ell = -2, -1, 0, +1, +2$

The electrons which make Zr different from Sr are d electrons, for which $\ell = 2$ (always). For $\ell = 2$, there are five values possible values of m_ℓ (-2, -1, 0, +1, +2 is five different numbers; see the concepts box at the beginning of the module if you do not understand how these values were derived), indicating that there are five different d orbitals

Spin Quantum Number

7. *What is the value of the spin quantum number, m_s, for the valence electrons in a Sr atom?*

The correct answer is: $m_s = +1/2$ and $-1/2$

Electronic Structure from the Periodic Chart

8. *What is the correct electronic structure of the Sr atom? Write the structure in both orbital notation and simplified (or spdf) notation.*

The correct answer is: [Kr] $\underset{\text{5s}}{\uparrow\downarrow}$ or [Kr] $5s^2$

s is the orbital angular momentum quantum number because Sr's valence electrons are in the s block of the periodic table.

This symbol indicates that 36 of the 38 electrons in Sr are in the same orbitals as in the noble gas Kr.

2 indicates that both of the distinguishing electrons in Sr are s electrons.

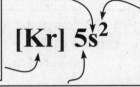

$[\text{Kr}]\ 5s^2$

5 is the principal quantum number because Sr is on the 5th row of the periodic table.

The $\uparrow$ indicates that the m_s for one of the valence electrons is $+1/2$. The $\downarrow$ symbolizes that $m_s = -1/2$ for the second valence electron.

As above, this symbol indicates that 36 of the 38 electrons in Sr are in the same orbitals as in the noble gas Kr.

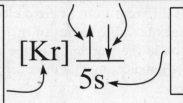

$[\text{Kr}]\ \underset{5s}{\uparrow\downarrow}$

5s is the symbol for the **n** and ℓ quantum numbers for Sr.

56

9. What is the correct electronic structure of the Zr atom? Write the structure in both orbital notation and simplified (or *spdf*) notation.

The correct answer is: $[Kr] \underset{5s}{\uparrow\downarrow} \;\; \underset{4d}{\uparrow \;\; \uparrow \;\; \_\_ \;\; \_\_ \;\; \_}$ or $[Kr]\, 5s^2\, 4d^2$

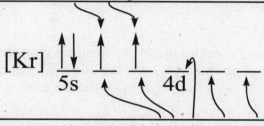

This symbol represents the two d electrons that differentiate Zr from Sr.

$$[Kr]\, 5s^2\, 4d^2$$

4d is the symbol for the **n** and **ℓ** quantum numbers for the d electrons in Zr.

These two arrows are not paired because of Hund's rule.

$[Kr] \underset{5s}{\uparrow\downarrow} \;\; \underset{\;\;\;\; 4d}{\uparrow \;\; \uparrow \;\; \_\_ \;\; \_\_ \;\; \_}$

The five spaces above the 4d symbol represent the five d orbitals that are possible in the 4d energy level.

Writing Quantum Numbers

10. Write the correct set of quantum numbers for the valence electrons in Sr.

The correct answer is:

n	ℓ	m_ℓ	m_s	
5	0	0	+1/2	1st electron quantum numbers
5	0	0	-1/2	2nd electron quantum numbers

INSIGHT: If both of the m_s numbers were reversed, the answer would still be correct.

11. *Write the correct set of quantum numbers for the valence electrons in Zr.*

The correct answer is:

n	ℓ	m_ℓ	m_s	
5	0	0	+1/2	1st electron quantum numbers
5	0	0	-1/2	2nd electron quantum numbers
4	2	-2	+1/2	3rd electron quantum numbers
4	2	-1	+1/2	4th electron quantum numbers

INSIGHT: Strictly speaking, m_ℓ could be any two of the five possible values and be correct.

CAUTION Both of the m_s values for the 4d electrons could also be -1/2 and be correct. But having one value +1/2 and the other -1/2 is incorrect because it does not obey Hund's rule.

YIELD There are several important rules that you need to know to understand electron configurations. These include ***Hund's rule, the Pauli Exclusion Principle, and the Aufbau Principle***. Be certain that you know and understand these rules. The most important thing to learn from this module is how to get the correct electronic configuration of an element using the periodic table.

Module 8 Predictor Questions

The following questions may help you to determine to what extent you need to study this module. The questions are ranked according to ability.

Leel 1 = basic proficiency
Leel 2 = mid level proficiency
Leel 3 = high proficiency

If you can correctly answer the Level 3 questions, then you probably do not need to spend much time with this module. If you are only able to answer the Level 1 problems, then you sould review the topics covered in this module.

Level 1
1. Select the element from each group that has the *largest* electronegativity
   a) S, Zn, Na, Te, Cu
   b) Al, Cr, Rb, Li, N
   c) K, Sb, Au, Cl, Ba
   d) Pd, Mg, O, Po, Sr

Level 1
2. Select the element with the *highest* first ionization energy.
   B, Al, Ga, In, Tl

Level 1
3. Which element has the *smallest* atomic radius?
   Mo, Au, Bi, In, Te

Level 3
4. Rank the following elements in order of *increasing* first ionization energy: C, B, N, O

Level 3
5. Which of the following elements has the *most negative* electron affinity? As, Al, K, Se, Sn

Level 2
6. Arrange these ions in order of *increasing* ionic radii.
   a) Al^{3+}, Na^+, Mg^{2+}
   b) F^-, N^{3-}, O^{2-}
   c) F^-, Na^+, Mg^{2+}, O^{2-}

Module 8 Predictor Question Solutions

1. Electronegativity increases up a group and from left to right across a period
   a) S
   b) N
   c) Cl
   d) O

2. First ionization energy increase up a group and from left to right across a priod.
   B has the greatest first ionization energy of these elements.

3. Atomic radius increases down a group and from right to left across a period so the atom in this group with the smallest atomic radius is Te.

4. The same trend described in number 2 is followed. However, N has a greater first ionization energy than O because its valence electron configuration contain 3 p electrons (half-filled p subshell). Therefore, N has a more energetically favorable configuration, and it is more difficult to remove an electron.
$$B < C < O < N$$

5. Electron affinity becomes more negative up a group and from left to right across a period. Thus, Se has the most negative electron affinity of this group.

6. When ions are isoelectric, ionic radius decreases with increasing atomic number. Keep in mind that cations have smaller radii than their neutral parent atoms while anions have larger radii than their neutral parents.

   a) $Al^{3+} < Mg^{2+} < Na^+$
   b) $F^- < O^{2-} < N^{3-}$
   c) $Mg^{2+} < Na^+ < F^- < O^{2-}$

Module 8
Chemical Periodicity

Introduction

There are many properties of elements that are based upon their electronic structures. Using some very simple rules, we can predict the variations of some properties based upon the element's position on the periodic chart. This module will help you learn the periodic properties that account for several important chemical properties. The primary goals of this module are to understand the periodic properties associated with:

1. electronegativity
2. ionization energy
3. electron affinity
4. atomic radii
5. ionic radii

You will need to have a periodic chart with you as you work on this module because you must learn to associate these properties on the periodic chart.

Module 8 Key Equations & Concepts

1. Electronegativity

Electronegativity is the relative measure of an element's ability to attract electrons to itself in a chemical compound. This property helps us determine the likelihood of ionic or covalent bond formation and the polarity of molecules.

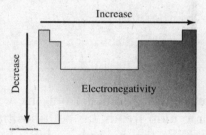

2. Ionization energy

Ionization energy is the amount of energy required to remove an electron from an atom or ion. This property is an important indicator of an element's likelihood of forming positive ions. Elements with several electrons can have a 1st ionization energy, 2nd ionization energy, and so forth until all of that element's electrons have been removed. Note the similarity to the trend in electronegativity.

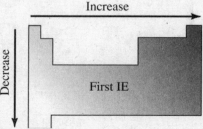

---

3. Electron affinity

Electron affinity is the amount of energy absorbed when an electron is added to an isolated gaseous atom. It will help us understand which elements are most likely to form negative ions. Electron affinity has the most irregular periodic trends of the properties discussed in this module.

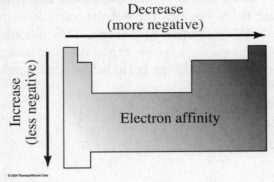

4. Atomic radii

Atomic radii are the measured distances from the center of the atom to its outer electrons. Atomic radii will help us predict the solid state structure of the elements.

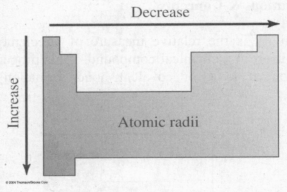

5. Ionic radii

This property is the measured distance from the center of an ion to its outer electrons. There are ionic radii trends for both positive and negative ions. These trends will help us determine the strength of ionic bonds. *Note that cations are always smaller than their parent atoms, and anions are always bigger than their parent atoms.*

IA	IIA		IIIA	IVA	VA	VIA	VIIA	VIIIA
Li^+ 0.90	Be^{2+} 0.59	Ionic radii			N^{3-} 1.71	O^{2-} 1.26	F^- 1.19	
Na^+ 1.16	Mg^{2+} 0.85		Al^{3+} 0.68			S^{2-} 1.70	Cl^- 1.67	
K^+ 1.52	Ca^{2+} 1.14		Ga^{3+} 0.76			Se^{2-} 1.84	Br^- 1.82	
Rb^+ 1.66	Sr^{2+} 1.32		In^{3+} 0.94			Te^{2-} 2.07	I^- 2.06	
Cs^+ 1.81	Ba^{2+} 1.49		Tl^{3+} 1.03				2 Å	

Electronegativity

1. Arrange these elements in the order of increasing electronegativity: O, Ca, Si, Cs
   The correct answer is: Cs < Ca < Si < O

| The most electronegative elements are in the upper right corner of the periodic chart. |

Electronegativities of the Elements

Metals	
Nonmetals	
Metalloids	

	IA																	VIIIA
1	1 H 2.1	IIA											IIIA	IVA	VA	VIA	VIIA	2 He
2	3 Li 1.0	4 Be 1.5											5 B 2.0	6 C 2.5	7 N 3.0	8 O 3.5	9 F 4.0	10 Ne
3	11 Na 1.0	12 Mg 1.2	IIIB	IVB	VB	VIB	VIIB		VIIIB		IB	IIB	13 Al 1.5	14 Si 1.8	15 P 2.1	16 S 2.5	17 Cl 3.0	18 Ar
4	19 K 0.9	20 Ca 1.0	21 Sc 1.3	22 Ti 1.4	23 V 1.5	24 Cr 1.6	25 Mn 1.6	26 Fe 1.7	27 Co 1.7	28 Ni 1.8	29 Cu 1.8	30 Zn 1.6	31 Ga 1.7	32 Ge 1.9	33 As 2.1	34 Se 2.4	35 Br 2.8	36 Kr
5	37 Rb 0.9	38 Sr 1.0	39 Y 1.2	40 Zr 1.3	41 Nb 1.5	42 Mo 1.6	43 Tc 1.7	44 Ru 1.8	45 Rh 1.8	46 Pd 1.8	47 Ag 1.6	48 Cd 1.6	49 In 1.6	50 Sn 1.8	51 Sb 1.9	52 Te 2.1	53 I 2.5	54 Xe
6	55 Cs 0.8	56 Ba 1.0	57 La 1.1 *	72 Hf 1.3	73 Ta 1.4	74 W 1.5	75 Re 1.7	76 Os 1.9	77 Ir 1.9	78 Pt 1.8	79 Au 1.9	80 Hg 1.7	81 Tl 1.6	82 Pb 1.7	83 Bi 1.8	84 Po 1.9	85 At 2.1	86 Rn
7	87 Fr 0.8	88 Ra 1.0	89 Ac 1.1 †															

*	58 Ce 1.1	59 Pr 1.1	60 Nd 1.1	61 Pm 1.1	62 Sm 1.1	63 Eu 1.1	64 Gd 1.1	65 Tb 1.1	66 Dy 1.1	67 Ho 1.1	68 Er 1.1	69 Tm 1.1	70 Yb 1.0	71 Lu 1.2
†	90 Th 1.2	91 Pa 1.3	92 U 1.5	93 Np 1.3	94 Pu 1.3	95 Am 1.3	96 Cm 1.3	97 Bk 1.3	98 Cf 1.3	99 Es 1.3	100 Fm 1.3	101 Md 1.3	102 No 1.3	103 Lr 1.5

© 2004 Thomson/Brooks Cole

| The least electronegative elements are in the lower left corner of the periodic chart. |

| Electronegativity steadily increases moving from the lower left to the upper right corners of the periodic chart. |

Ionization Energy

2. Arrange these elements by increasing ionization energies: F, N, C, O
   The correct answer is: C < O < N < F

| First ionization energies increase steadily from the alkali metals to the noble gases. |

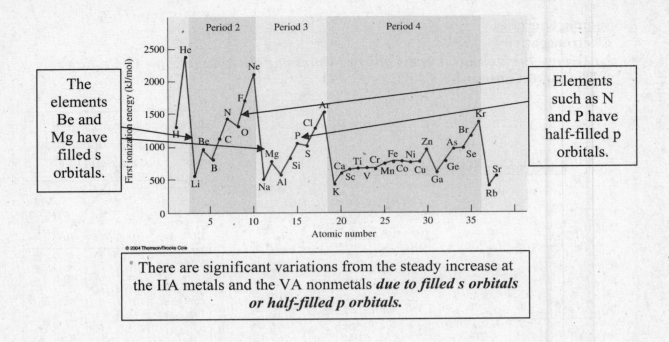

The elements Be and Mg have filled s orbitals.

Elements such as N and P have half-filled p orbitals.

© 2004 Thomson/Brooks Cole

There are significant variations from the steady increase at the IIA metals and the VA nonmetals *due to filled s orbitals or half-filled p orbitals.*

Electron Affinity

3. Arrange these elements by increasing electron affinity: F, N, C, O
    The correct answer is: F < O < C < N

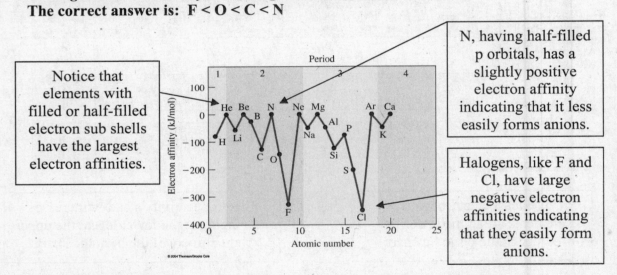

Notice that elements with filled or half-filled electron sub shells have the largest electron affinities.

N, having half-filled p orbitals, has a slightly positive electron affinity indicating that it less easily forms anions.

Halogens, like F and Cl, have large negative electron affinities indicating that they easily form anions.

© 2004 Thomson/Brooks Cole

The generic trend shown in the concepts box does not work in comparing the electron affinities of C and N. N is farther to the left in the same row as C, but because N has a half-filled p orbital its electron affinity is slightly positive and greater than that of C.

Atomic Radii

4. *Arrange these elements by increasing atomic radii: F, Ga, S, Rb*
    The correct answer is: F < S < Ga < Rb

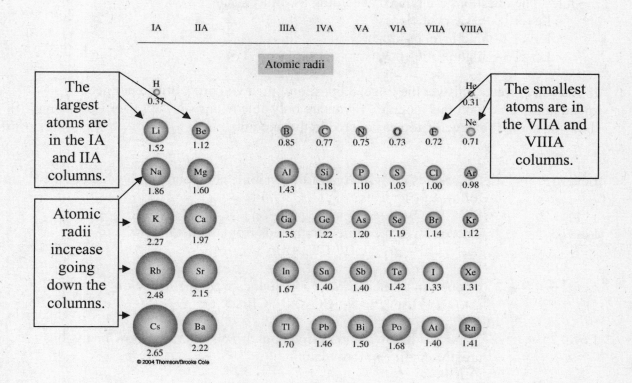

| | IA | IIA | | IIIA | IVA | VA | VIA | VIIA | VIIIA |

Atomic radii

The largest atoms are in the IA and IIA columns.

Atomic radii increase going down the columns.

The smallest atoms are in the VIIA and VIIIA columns.

H 0.37
He 0.31

Li 1.52
Be 1.12
B 0.85
C 0.77
N 0.75
O 0.73
F 0.72
Ne 0.71

Na 1.86
Mg 1.60
Al 1.43
Si 1.18
P 1.10
S 1.03
Cl 1.00
Ar 0.98

K 2.27
Ca 1.97
Ga 1.35
Ge 1.22
As 1.20
Se 1.19
Br 1.14
Kr 1.12

Rb 2.48
Sr 2.15
In 1.67
Sn 1.40
Sb 1.40
Te 1.42
I 1.33
Xe 1.31

Cs 2.65
Ba 2.22
Tl 1.70
Pb 1.46
Bi 1.50
Po 1.68
At 1.40
Rn 1.41

© 2004 Thomson/Brooks Cole

Ionic Radii

5. *Arrange these ions by increasing ionic radii: S^{2-}, Cl^-, Mg^{2+}, Al^{3+}*
    The correct answer is $Al^{3+} < Mg^{2+} < Cl^- < S^{2-}$

| IA | IIA | | IIIA | IVA | VA | VIA | VIIA | VIIIA |

Ionic radii

Positive ions are smaller than their atoms. The more positive the ion, the smaller its radius.

Negative ions are larger than their atoms. The more negative the ion, the larger it is.

Like atomic radii, ionic radii increase going down the columns.

Li^+ 0.90
Be^{2+} 0.59

N^{3-} 1.71
O^{2-} 1.26
F^- 1.19

Na^+ 1.16
Mg^{2+} 0.85
Al^{3+} 0.68

S^{2-} 1.70
Cl^- 1.67

K^+ 1.52
Ca^{2+} 1.14
Ga^{3+} 0.76

Se^{2-} 1.84
Br^- 1.82

Rb^+ 1.66
Sr^{2+} 1.32
In^{3+} 0.94

Te^{2-} 2.07
I^- 2.06

Cs^+ 1.81
Ba^{2+} 1.49
Tl^{3+} 1.03

2 Å

© 2004 Thomson/Brooks Cole

Module 9 Predictor Questions

The following questions may help you to determine to what extent you need to study this module. The questions are ranked according to ability.

Level 1 = basic proficiency
Level 2 = mid level proficiency
Level 3 = high proficiency

If you can correctly answer the Level 3 questions, then you probably do not need to spend much time with this module. If you are only able to answer the Level 1 problems, then you should review the topics covered in this module.

Level 1   1. Chlorine is **most likely** to form an ionic compound with which of the following elements? F, O, C, N, Li

Level 3   2. Choose all of the **ionic** compounds from the list below:
          K_3N, $CaBr_2$, Li_2O, HI, CF_4, OBr_2

Level 3   3. Choose all of the **covalent** compounds from the list below:
          $Ca(OH)_2$, Li_3N, Sr_3N_2, CO_2, NI_3, CBr_4

Level 2   4. Name the following ionic compounds and determine how many ions are present in one formula unit.
          a) $AlPO_4$
          b) $Mg(NO_3)_2$
          c) Na_2CO_3

Level 1   5. Draw the Lewis dot structures of the following atoms: B, P, K, and S

Level 1   6. Which of the following formulas is **incorrect**?
          $SrBr_2$, K_2S, MgSe, $CsCl_2$, Al_2O_3

Level 1   7. Which one of the following compounds involves **both** ionic and covalent bonding? Cl_2, Na_2SO_4, KCl, HF, HCN

Level 1   8. Which one of the following molecules **does not** have a dipole moment (which is nonpolar?) BrCl, ClF, BrF, O_2, ICl

Module 9 Predictor Question Solutions

1. Chlorine is a nonmetal that forms the anion Cl^- It is most likely to form an ionic compound with a metal such as Li (forms the Li^+ cation).

2. Ionic compounds are formed between a metal and a nonmetal OR a metal and polyatomic anion. The ionic compounds in this list are: K_3N, $CaBr_2$, and Li_2O. Note that HI is NOT ionic since H and I are both nonmetals.

3. Covalent compounds from between two or more nonmetals. The covalent compounds in this list are: CO_2, NI_3, and CBr_4.

4. a) $AlPO_4$ is aluminum phosphate; two ions
   b) $Mg(NO_3)_2$; magnesium nitrate; three ions
   c) Na_2CO_3; sodium carbonate; three ions

5.

   B·   ·P·   K   ·S:

6. $CsCl_2$ is the incorrect formula. Cs forms a 1+ cation while Cl forms a 1- anion. The correct formula is CsCl.

7. Ionic compounds form between a metal and a nonmetal OR a metal and a polyatomic ion. Covalent compounds form between two or more nonmetals. In this list, Na_2SO_4 is the only molecule that contains both types of bonding. The bond between the two Na^+ ions and the SO_4^{2-} polyatomic anion is ionic in nature. However, the bonds that hold SO_4^{2-} together are covalent since S and O are both nonmetals.

8. O_2 does not contain a dipole. Dipoles result from the unequal sharing of an electron pair in a covalent bond. The unequal sharing is the result of the two atoms in the bond having different electronegativities. Since O_2 contains O bound to O, there is no difference in electronegativity and no dipole

Module 9
Chemical Bonding

Introduction

This module explains how chemical bonds are formed. There are two basic types of chemical bonds: ionic and covalent. This module's goals include:

1. learning to determine if a compound is ionic or covalent,
2. drawing Lewis dot structures of atoms
3. writing formulas of the simple ionic compounds
4. determining relative ionic bond strength
5. drawing Lewis dot structures of ionic and covalent compounds
6. recognizing if a covalent bond is polar or nonpolar

A periodic table will help you understand many of the electron structures used in this chapter.

Module 9 Key Equations & Concepts

$$\text{Force of attraction between 2 ions} \propto \frac{q^+ \times q^-}{d^2}$$

$$\text{Force of attraction between 2 ions} \propto \frac{(\text{Charge on cation})(\text{Charge on anion})}{(\text{Distance between the ions})^2}$$

Coulomb's Law describes the strength of attraction between two ions of opposite charge. This can be used to determine the strength of ionic bonds.

Sample Exercise

Determining if a Compound is Ionic or Covalent

1. *Indicate which of the following compounds are ionic in nature and which are covalent in nature.*

$$CH_4, \ KBr, \ Ca_3N_2, \ Cl_2O_7, \ H_2SO_4, \ InCl_3$$

The correct answer is: ionic = KBr, Ca_3N_2 and $InCl_3$
covalent = CH_4, Cl_2O_7, and H_2SO_4

 Look for metallic elements! Ionic compounds are formed by the reaction of metallic elements with nonmetallic elements or the reaction of the ammonium ion, NH_4^+, with nonmetals. Covalent compounds are formed by the reaction of two or more nonmetals

K is a metallic element.	Ca is a metallic element.	In is a metallic element.
KBr	**Ca₃N₂**	**InCl₃**
Br is a nonmetal.	N is a nonmetal	Cl is a nonmetal

C is a nonmetal.	Cl is a nonmetal.	H and S are nonmetals.
CH$_4$	**Cl$_2$O$_7$**	**H$_2$SO$_4$**
H is a nonmetal.	O is a nonmetal.	O is a nonmetal.

Lewis Dot Structures of Atoms Sample Exercises

2. *Draw the correct Lewis dot structure of these elements: Mg, P, S, Ar*
   The correct Lewis dot structures are shown below.

 TIPS Use the periodic table to determine how many valence electrons from each element's group number! Your first step in drawing a Lewis dot structure should *always* be to carry out an electron count for the species.

The number next to each dot represents the order in which it was added to the structure. Essentially, each of the four sides of the element's symbol represents an orbital. One is a s orbital, and the remaining three are p orbitals. The s orbital must be filled first, followed by the three p orbitals. Note that it does not matter where you start or whether you proceed clockwise or counterclockwise, as long as you follow Hund's rule and the Aufbau principle.

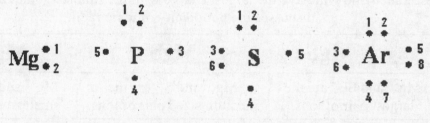

[Ne] ↑↓ 3s	[Ne] ↓↑ ↑ ↑ ↑ 3s 3p	[Ne] ↓↑ ↑↓ ↑ ↑ 3s 3p	[Ne] ↓↑ ↑↓ ↑↓ ↑↓ 3s 3p

> Lewis dot structures reflect the electronic structures of the elements, including how the electrons are paired. Notice how the orbital diagrams match the Lewis dot structures of each element.

Simple Ionic Compounds

3. *Write the correct formulas of the ionic compounds formed when Mg atoms react with the following: a) Cl atoms, b) S atoms, c) P atoms.*
   The correct answers are: MgCl$_2$, MgS, and Mg$_3$P$_2$

> Like all of the IIA metals, Mg has two electrons in its valence shell and commonly forms 2+ ions, Mg^{2+}.

MgCl₂	**MgS**	**Mg₃P₂**
Cl, and all of the VIIA nonmetals, have seven electrons in their valence shell and commonly form 1- ions, Cl⁻. Two Cl⁻ ions are required to balance the 2+ charge of the Mg and form a neutral compound.	S, and all of the VIA nonmetals, have six electrons in their valence shell and commonly form 2- ions, S^{2-}. Only one S^{2-} ion is required to cancel out the 2+ charge on the Mg^{2+} ion.	P, and all of the VA nonmetals, have five electrons in their valence shell and commonly form 3- ions, P^{3-}. Two P^{3-} ions are needed to balance the charge on three Mg^{2+} ions to form a neutral compound.

4. *Arrange these ionic compounds by increasing strength of the ionic bond in each compound: MgSe, MgO, MgS*
    The correct answer is: MgSe < MgS < MgO

> Coulomb's Law tells us that the force of attraction between ions $= \dfrac{q^{+} \times q^{-}}{d^2}$. The strongest ionic bond will have the largest charge with the smallest ionic radii. Module 8 discusses the periodicity of ionic radii.

MgSe   <   MgS   <   MgO

Mg^{2+} and Se^{1-} are the largest pair of ions.	Mg^{2+} and S^{2-} are the medium sized pair of ions.	Mg^{2+} and O^{2-} are the smallest pair of ions.

△ TIP: When comparing the strength of ionic bonds in compounds that all contain a common cation, simply compare the ionic radii of the anions. The smaller the anion, the greater the strength of the bond. The same method can be used if the compounds contain a common anion and the cations vary.

Drawing Lewis Dot Structures of Ionic Compounds
5. *Draw the Lewis dot structures for each of these compounds: AlP, NaCl, MgCl₂*
    The correct structures are shown below.

When counting the valence electrons, remember that a cation has, per positive charge, one electron less than the neutral parent atom. Per each negative charge, anions have one electron more than the neutral parent atom.

Al^{3+} $\left[:\ddot{P}: \right]^{3-}$ Na^{+} $\left[:\ddot{Cl}: \right]^{-}$ $Mg^{2+}\ 2\left[:\ddot{Cl}: \right]^{-}$

Al loses all three of its valence electrons and forms a 3+ ion. Thus, it has no dots.

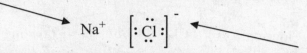

$Al^{3+} \left[:\ddot{P}: \right]^{3-}$

P gains three electrons from Al, so it has 8 dots (5 valence electron plus 3 from Al) and forms a 3- ion. The []'s indicate that the 3- charge is associated with the P ion.

Na loses its one valence electron to form a 1+ ion, so it has no dots.

$Na^{+} \left[:\ddot{C}l: \right]^{-}$

Cl gains one electron from Na, thus has 8 dots (7 valence electron plus 1 from Na), and forms a 1- ion.

Mg loses both valence electrons in forming a 2+ ion, so it has no dots.

$Mg^{2+} \, 2 \left[:\ddot{C}l: \right]^{-}$

Each Cl atom gains one electron from the Mg. The 2 in front of the []'s indicates that two Cl⁻ ions are needed to balance the charge of the Mg^{2+} ion.

Simple Covalent Compounds

6. Draw the correct Lewis dot structures for each of these compounds: SiH_4, PCl_3, SF_6 The correct structures are:

$$
\begin{array}{c}
H \\
H:\ddot{S}i:H \\
H
\end{array}
\qquad
\begin{array}{c}
:\ddot{C}l:\ddot{P}:\ddot{C}l: \\
:\ddot{C}l:
\end{array}
\qquad
\begin{array}{ccc}
 & \ddot{F} & \\
F & F & F \\
 & S & \\
F & F & F
\end{array}
$$

Try following these steps when drawing Lewis structures:
1. Determine the number of valence electrons in the compound
2. Decide which atom is the central atom and make one bond (two electrons) to each of the remaining elements.
3. Fill in the octet for all elements, and count how many electrons have been used. Procedures to apply when there are too many or too few electrons will be discussed later in this module.

SiH$_4$ has 8 valence electrons (4 from Si and 1 from each of the 4 H)

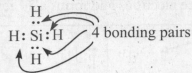

4 bonding pairs

In most cases every element in a compound will obey the octet rule. Thus, Si has a share of 8 electrons and each H has a share of 2 electrons. This compound has only _bonding pairs_ of electrons.

PCl$_3$ has 26 valence electrons (5 from P and 7 from each of the 3 Cl).

lone pair

3 bonding pairs

In this compound P has a share of 8 electrons and each Cl has a share of 8 electrons. This compound has 3 _bonding pairs_ and 1 _lone pair_ of electrons.

SF$_6$ has 48 valence electrons (6 from S and 7 from each of the 6 F).

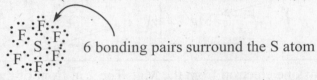

6 bonding pairs surround the S atom

This compound does not obey the octet rule.

S has a share of 12 electrons while each F has a share of 8 electrons. This compound has 6 _bonding pairs_ of electrons. Look in your textbook for the rules on which compounds do not obey the octet rule.

INSIGHT: When drawing Lewis dot structures, if the compound obeys the octet rule, the central atom will have a share of 8 electrons. The possible combinations of 8 electrons for compounds that **obey the octet rule** are:

Bonding Pairs	Lone Pairs
4	0
3	1
2	2
1	3

INSIGHT: If the compound **does not obey the octet rule**, the central atom can have 2, 3, 5 or 6 pairs of electrons around the central atom in some combination of bonding and lone pairs.

INSIGHT: Noncentral atoms will obey the octet rule having either 1 bonding pair, as for H atoms, or a share of 8 electrons as is the case for Cl and F in the examples above.

Compounds Containing Multiple Bonds

7. Draw the correct Lewis dot structures for each of these compounds: CO_2 and N_2
   The correct structures are:

$$:\ddot{O} :: C :: \ddot{O}:$$

$$:N ::: N:$$

CO_2 has 16 valence electrons (4 from C and 6 from each of the 2 O).
C will be the central atom. Connecting each O with the central C by one bonding pair and filling in all octets results in the following structure:

$$:\ddot{O} \cdot\cdot \ddot{C} \cdot\cdot \ddot{O}:$$

Note that this structure contains 20 electrons, which is four more than the structure should have.

 TIP In order to decrease the number of electrons in a Lewis structure, make a double bond. *When the double bond is made one lone pair of electrons must be removed from each atom involved in the double bond.*

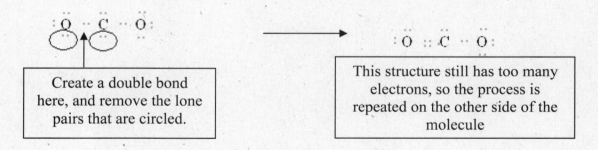

Create a double bond here, and remove the lone pairs that are circled.

This structure still has too many electrons, so the process is repeated on the other side of the molecule

N_2 has 10 valence electrons. Connecting the two atoms and filling each octet results in a structure with 14 electrons:

$$:\ddot{N} \cdot\cdot \ddot{N}:$$

The formation of each multiple bond reduces the total electron count by 2 electrons. In this case, the process must be carried out twice in order to remove four electrons. The result is a *triple bond*.

Polar or Nonpolar Covalent Bonds in Compounds
8. Which of these compounds contains polar covalent bonds?
$$F_2, CH_4, H_2O$$

The correct answer is: CH_4 and H_2O contain polar covalent bonds and F_2 does not

The periodic trends regarding electronegativity are discussed in Module 8. You will need to know those trends for problems of this nature.

Polar covalent bonds occur when the two atoms involved in the bond have a difference in electronegativity. In F_2 the two atoms are both F. They have the same electronegativity; thus, there is not a polar bond. In CH_4 and H_2O, the H to the central atom (C or O) bond involves atoms with different electronegativities. Thus there are polar covalent bonds in CH_4 and H_2O.

INSIGHT:
> Polar bonds have **dipoles** resulting from the *partial positive* and *partial negative* charges on atoms resulting from the sharing of electrons. Dipoles are indicated by drawing an arrow over the bond with the head of the arrow pointing in the direction of the more electronegative atom. Each C-O bond in CO_2 is polar, as shown below:
>
> $$\overset{\longleftarrow \quad \longrightarrow}{\ddot{O} :: C :: \ddot{O}}$$

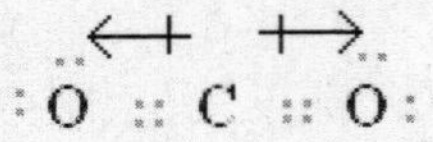

Module 10 Predictor Questions

The following questions may help you to determine to what extent you need to study this module. The questions are ranked according to ability.

Level 1 = basic proficiency
Level 2 = mid level proficiency
Level 3 = high proficiency

If you can correctly answer the Level 3 questions, then you probably do not need to spend much time with this module. If you are only able to answer the Level 1 problems, then you should review the topics covered in this module.

Level 1   1. Determine the electronic and molecular geometries of each molecule from its Lewis structure.
a) CBr_4
b) F_2
c) SF_6
d) PCl_5
e) BF_3

Level 2   2. What is the molecular geometry of PCl_3?

Level 3   3. Which species is *incorrectly* matched with bond angles?

Molecule	Bond Angles
HCN	180°
ClO_3^-	slightly < 109°
NH_3	107°
SeO_4^{2-}	109.5°
CCl_4	90°, 120°, and 180°

Module 10 Predictor Question Solutions

1.

Example	Electronic Geometry	Molecular Geometry
a)	Tetrahedral	Tetrahedral
b)	Linear	Linear
c)	Octahedral	Octahedral
d)	Trigonal bipyramidal	Trigonal bipyramidal
e)	Trigonal planar	Trigonal planar

2. The Lewis structure for PCl_3 should contain 26 valence electrons. P is the central atom with three single bonds to Cl atoms and one lone pair. Thus, the electronic geometry is tetrahedral, but the presence of the lone pair results in **trigonal pyramidal molecular geometry**.

3. HCN has linear electronic and molecular geometry, so it has 180° bond angles.

ClO_3^- has tetrahedral electronic geometry, but a lone pair on the central Cl atom results in trigonal pyramidal molecular geometry and bond angles slightly less than the standard 109° for tetrahedral molecules.

NH_3 also has tetrahedral electronic geometry but a lone pair on the central N atom results in trigonal pyramidal molecular geometry and bond angles slightly less than the standard 109° for tetrahedral molecules.

SeO_4^{2-} has tetrahedral electronic and molecular geometry. Its bond angles are the standard 109°.

CCl_4 has tetrahedral electronic and molecular geometry. Its bond angles are the standard 109° rather than the 90°, 120°, and 180° listed.

Module 10
Molecular Shapes

Introduction

Molecular shape refers to the geometrical arrangement of atoms around the central atom in a molecule or polyatomic ion. This module will help you understand and predict the stereochemistry of many molecules. Molecular shapes are important in the chemical reactivity of numerous compounds. The most important goal of this module is to learn to:

1. predict and names, electronic geometries, and molecular geometries of molecules

---

Module 10 Key Equations & Concepts

All of the molecules described in this module have two geometries that you must be familiar with, their electronic and molecular geometries.

1) **Electronic geometry** considers all of the regions of high electron density including bonding pairs, lone pairs, and double or triple bonds.

2) **Molecular geometry** only considers those electrons and atoms that are involved in bonding pairs or in double and triple bonds.

The molecular geometry will be different from the electronic geometry only in molecules that have lone pairs of electrons.

Regions of electron density	Electronic geometry	Molecular geometry	Bond Angles	Examples
2	Linear	Linear	180°	BeF_2, BeH_2, $BeCl_2$
3	Trigonal planar	Trigonal planar	120°	BH_3, $AlCl_3$, BF_3
4	Tetrahedral	Tetrahedral, trigonal pyramidal, or linear	Vary	CH_4, SiH_4, PF_3, H_2O
5	Trigonal bipyramidal	Trigonal bipyramidal, see-saw, T-shaped, or linear	Vary	PF_5, SF_4, ClF_3, XeF_2
6	Octahedral	Octahedral, square pyramidal, or square planar	Vary	SF_6, IF_5, XeF_4

CAUTION

Note that geometries are based on the number of high electron density regions (both bonding and nonbonding pairs) *around the central atom.* Remember that all types of bonds (single, double, and triple) count as ONE region of electron density.

It is impossible to determine the correct geometry of a molecule if you do not start with the correct Lewis structure!

Number of Regions of High Electron Density	Description; Angles†	Electronic Geometry*	
		Line Drawing‡	Ball-and-stick model
2	linear; 180°		
3	trigonal planar; 120°		
4	tetrahedral; 109.5°		
5	trigonal bipyramidal; 90°, 120°, 180°		
6	octahedral; 90°, 180°		

© 2004 Thomson/Brooks Cole

Sample Exercises

Linear Molecules
1. *What are the correct molecular geometries of BeI$_2$ and BeHF?*
   The correct answer is: Both molecules have linear molecular geometries.

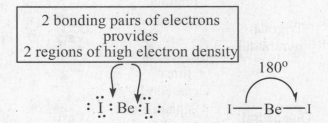

2 bonding pairs of electrons
provides
2 regions of high electron density

180°

Notice that the linear shape is determined by the electrons around the central Be atom, not the lone pairs on the I atoms.

78

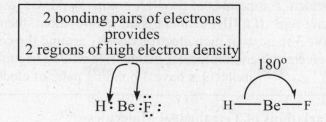

2 bonding pairs of electrons provides
2 regions of high electron density

180°

The linear shape is not affected by the two different atoms bonded to Be. The shape is determined by the 2 regions of high electron density.

INSIGHT: Covalent Compounds of Be do not obey the octet rule. If Be is the central atom in a molecule there will be 2 regions of high electron density and the electronic and molecular geometries will be linear.

Trigonal Planar Molecules

2. *What are the correct molecular geometries of BH_3 and AlHFBr?*
   The correct answer is: Both molecules have trigonal planar molecular geometries.

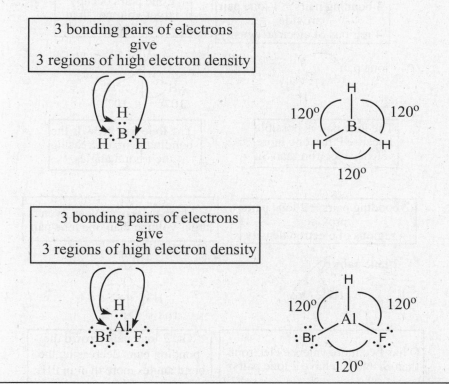

3 bonding pairs of electrons give
3 regions of high electron density

120°   120°

120°

3 bonding pairs of electrons give
3 regions of high electron density

120°   120°

120°

Again, having 3 different atoms does not affect the molecule's shape. Shapes are determined by the regions of high electron density.

79

Covalent compounds of the IIIA group (B, Al, Ga, & In) do not obey the octet rule. If a IIIA element is the central atom, then the molecule will have 3 regions of high electron density around the central atom and the electronic and molecular geometries will be trigonal planar. These molecules have 3 bonding pairs of electrons.

Tetrahedral and Variations of Tetrahedral Molecules
3. *What are the correct molecular geometries of SiH$_4$, PF$_3$, and H$_2$O?*
   The correct answers are: tetrahedral, trigonal pyramidal, and bent, respectively.

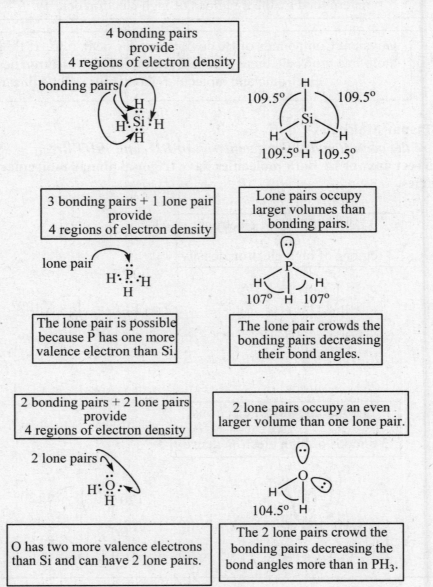

All of the molecules in this category will obey the octet rule because they have 4 regions of high electron density which is equivalent to 8 electrons.

1. If a IVA element (C, Si, or Ge) is the central atom, the electronic and molecular geometries will be tetrahedral. These molecules contain 4 bonding pairs of electrons.
2. If a VA element (N, P, or As) is the central atom, the electronic geometry will be tetrahedral and the molecular geometry will be trigonal pyramidal. These molecules contain 3 bonding pairs and 1 lone pair of electrons.
3. If a VIA element (O, S, Se) is the central atom, the electronic geometry will be tetrahedral and the molecular geometry will be bent, angular, or V-shaped. These molecules contain 2 bonding pairs and 2 lone pairs of electrons.
4. If a VIIA element (F, Cl, Br, or I) is the central atom, the electronic geometry will be tetrahedral and the molecular geometry will be linear. These molecules contain 1 bonding pair and 3 lone pairs of electrons.

TIP When the central atom has no lone pairs, the molecular and electronic geometries are the same.

Trigonal Bipyramidal and Variations of Trigonal Bipyramidal Molecules
4. *What are the correct molecular geometries of PF_5, SF_4, ClF_3, and XeF_2*
   The correct answer is trigonal bipyramidal for PF_5, see-saw shaped for SF_4, T-shaped for ClF_3, and linear for XeF_2.

5 bonding pairs on P atom provide
5 regions of electron density

| 4 bonding pairs + 1 lone pair provide 5 regions of electron density | The increased volume of the lone pair on S changes the bond angles between the F atoms. |

lone pair

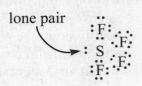

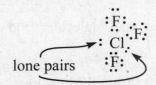

177°   102°   90°

| S has one more valence electron than P which makes the lone pair. | The see-saw shape is a simple modification of trigonal bipyramid due to the lone pair. |

| 3 bonding pairs + 2 lone pairs provide 5 regions of electron density | Notice that both lone pairs occupy equatorial positions. |

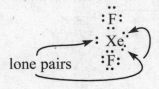

lone pairs

90°   180°

| Cl has two more valence electrons than P which makes the two lone pairs. | The T- shape is another modification of trigonal bipyramid due to two lone pairs. |

| 2 bonding pairs + 3 lone pairs provide 5 regions of electron density | Notice that all three lone pairs occupy equatorial positions. |

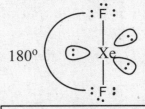

lone pairs

180°

| Xe has three more valence electrons than P which makes the three lone pairs. | This linearshape is another modification of trigonal bipyramid due to three lone pairs. |

None of the molecules in this category obey the octet rule. They will have 5 regions of high electron density or 10 total electrons around the central atom.

1. If a VA element (P or As) is the central atom, the electronic and molecular geometries will be trigonal bipyramidal. These molecules contain 5 bonding pairs of electrons.

2. If a VIA element (S or Se) is the central atom, the electronic geometry will be trigonal bipyramidal and the molecular geometry will be seesaw. These molecules contain 4 bonding pairs and 1 lone pair of electrons.

3. If a VIIA element (Cl, Br, or I) is the central atom, the electronic geometry will be trigonal bipyramidal and the molecular geometry will be T-shaped. These molecules contain 3 bonding pairs and 2 lone pairs of electrons.

4. If an VIIIA element (Xe or Kr) is the central atom, the electronic geometry will be trigonal bipyramidal and the molecular geometry will be linear. These molecules contain 2 bonding pairs and 3 lone pairs of electrons.

Octahedral and Variations of Octahedral Molecules

5. *What are the correct molecular geometries of SF_6, IF_5, and XeF_4?*
   The correct answers are: octahedral for SF_6, square pyramidal for IF_5, and square planar for XeF_4.

> 6 bonding pairs on S
> provides
> 6 regions of electron density

5 bonding pairs and 1 lone pair provides 6 regions of electron density	The square-based pyramid shape is the octahedral shape with one lone pair of electrons.

90°

lone pair

I has one more valence electron than S making the lone pair possible.

4 bonding pairs and 2 lone pairs provides 6 regions of electron density	The square planar shape is the octahedral shape with two lone pairs of electrons.

lone pairs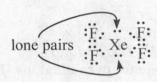

90°

Xe has two more valence electrons than S making the two lone pairs possible.

INSIGHT:

None of the molecules in this category obey the octet rule. They will have 6 regions of high electron density or 12 total electrons around the central atom.

1. If a VIA element (S or Se) is the central atom, the electronic and the molecular geometries will be octahedral. These molecules contain 6 bonding pairs of electrons.

2. If a VIIA element (Cl, Br, or I) is the central atom, the electronic geometry will be octahedral and the molecular geometry will be square pyramidal. These molecules contain 5 bonding pairs and 1 lone pair of electrons.

3. If an VIIIA element (Xe or Kr) is the central atom, the electronic geometry will be octahedral and the molecular geometry will be square planar. These molecules contain 4 bonding pairs and 2 lone pairs of electrons.

Module 11 Predictor Questions

The following questions may help you to determine to what extent you need to study this module. The questions are ranked according to ability.

Level 1 = basic proficiency
Level 2 = mid level proficiency
Level 3 = high proficiency

If you can correctly answer the Level 3 questions, then you probably do not need to spend much time with this module. If you are only able to answer the Level 1 problems, then you should review the topics covered in this module.

Level 1   1. What kind of hybrid orbitals are utilized by the carbon atom in CF_4 molecules?

Level 1   2. According to valence bond theory, what is the hybridization at the sulfur atom in SF_6?

Level 2   3. What is the hybridization of a carbon atom involved in a triple bond?

Level 2   4. Show the dipoles for the following molecules.
CO_2, NI_3, OF_2, CH_2Cl_2

Level 1   5. Determine which of the following molecules is nonpolar.
CCl_4, CH_2Cl_2, CH_3Cl, $CHCl_3$, SiH_2Cl_2

Level 3   6. Which one of the following is a *nonpolar* molecule with *polar* covalent bonds? NH_3, H_2Te, $SOCl_2$ (S is the central atom), $BeBr_2$, HF

Module 11 Predictor Question Solutions

1. CF_4 is a tetrahedral molecule containing four C-F single bonds. There are therefore four electron groups surrounding the central C atom, and this correlates to sp^3 hybridization.

2. SF_6 is an octahedral molecule containing six S-F single bonds. There are therefore six electron groups surrounding the central S atom, and this correlates to sp^3d^2 hybridization.

3. Carbon forms four bonds, so if it is engaged in a triple bond then it can have only two electron groups (one from the triple bond, one from the remaining single bond). This correlates to sp hybridization.

4.

5. In symmetrical molecules, bond dipoles may cancel one another. The nonpolar molecule in this list is CCl_4.

6. In polar molecules dipoles do not cancel. All of the molecules in this list contain polar bonds, but only in $BeBr_2$ (with linear geometry) do the dipoles cancel resulting in a nonpolar molecule.

Module 11
Hybridization and Polarity of Molecules

Introduction

Valence Bond theory is another way of describing the shapes of molecules. It involves the hybridization (mixing) of atomic orbitals. The names of the orbitals come from the orbitals that have been mixed to make the shape. This module will help you understand and predict the hybridization of the atoms in several molecules. Polarity refers to whether the electron density of a molecule is symmetrically or asymmetrically arranged about the molecule. The goals of this module are:

1. to learn to predict the hybridization of atoms using Valence Bond theory
2. to understand the hybridization and geometry of double and triple bonds
3. to learn how to determine the polarity of a molecule

Module 11 Key Equations & Concepts

1. **sp hybridized atoms**
   Atoms having two regions of electron density and a linear electronic geometry can be described as having orbitals that are made from one s and one p orbital.

2. **sp^2 hybridized atoms**
   Atoms having three regions of electron density and a trigonal planar electronic geometry can be described as having orbitals that are made from one s and two p orbitals.

3. **sp^3 hybridized atoms**
   Atoms having four regions of electron density and a tetrahedral electronic geometry can be described as having orbitals that are made from one s and three p orbitals.

4. **sp^3d hybridized atoms**
   Atoms having five regions of electron density and a trigonal bipyramidal electronic geometry can be described as having orbitals that are made from one s, three p, and one d orbitals.

5. **sp^3d^2 hybridized atoms**
   Atoms having six regions of electron density and an octahedral electronic geometry can be described as having orbitals that are made from one s, three p, and two d orbitals.

The number of regions of electron density describes both the electronic geometry and the hybridization.

Regions of Electron Density	Electronic Geometry	Hybridization
2	Linear	sp
3	Trigonal planar	sp^2
4	Tetrahedral	sp^3
5	Trigonal bipyramidal	sp^3d
6	Octahedral	sp^3d^2

TIP The key to answering questions involving geometry and hybridization in compounds containing double and triple bonds is counting the regions of high electron density surrounding the atom in question. Double and triple bonds count as one region of electron density. Lone pairs also count as one region of electron density.

INSIGHT: The number of hybrid orbitals formed is equal to the number of atomic orbitals that were combined. For example:
One s orbital + one p orbital = 2 sp hybrid orbitals

Sample Exercises
Hybridization
1. What is the hybridization of the underlined atom in each of the following molecules?
$\underline{Be}I_2$, $\underline{B}H_3$, $\underline{Si}H_4$, $\underline{P}F_5$, $\underline{S}F_6$

The correct answers are: sp, sp^2, sp^3, sp^{3}d, and sp^{3}d^2, respectively,

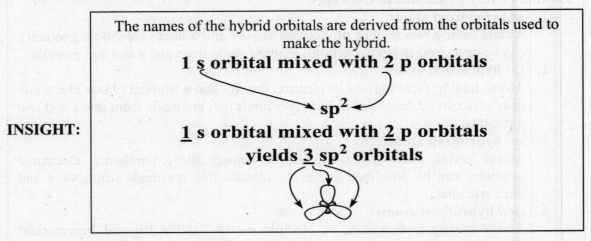

INSIGHT: The names of the hybrid orbitals are derived from the orbitals used to make the hybrid.
1 s orbital mixed with 2 p orbitals
sp^2
1 s orbital mixed with 2 p orbitals
yields 3 sp^2 orbitals

$\underline{Be}I_2$

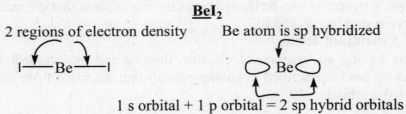

2 regions of electron density   Be atom is sp hybridized

I—Be—I   Be

1 s orbital + 1 p orbital = 2 sp hybrid orbitals

INSIGHT: Any central atom that has **2 regions of high electron density and a linear electronic geometry** can be described as an **sp hybrid**. See Sample Exercises 2 and 3 below for examples of different central atoms.

BH₃

3 regions of electron density   B atom is sp² hybridized

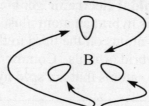

1 s orbital + 2 p orbitals = 3 sp² hybrid orbitals

INSIGHT: Any central atom that has **3 regions of high electron density and a trigonal planar electronic geometry** can be described as an **sp² hybrid**. See Sample Exercises 2 and 3 below for examples of different central atoms.

SiH₄

4 regions of electron density   Si atom is sp³ hybridized

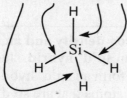

1 s orbital + 3 p orbitals = 4 sp³ hybrid orbitals

INSIGHT: Any central atom that has **4 regions of high electron density and a tetrahedral electronic geometry** can be described as an **sp³ hybrid**. See Sample Exercises 2 and 3 below for examples of different central atoms.

PF₅

5 regions of electron density   P atom is sp³d hybridized

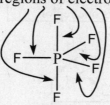

1 s orbital + 3 p orbitals + 1 d orbital = 5 sp³d hybrid orbitals

$\underline{SF_6}$

6 regions of electron density   S atom is sp³d² hybridized

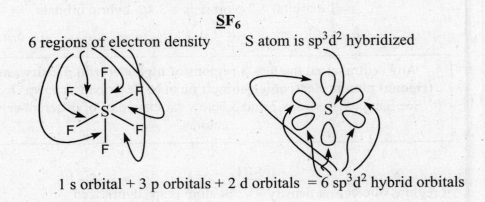

1 s orbital + 3 p orbitals + 2 d orbitals = 6 sp³d² hybrid orbitals

Hybridization of Double and Triple Bonds

2. *What is the hybridization of the underlined atoms in these molecules?*
$$\underline{C_2H_4}, \ \underline{C_2H_2}, \ H_2\underline{C}O$$

The correct answers are: sp², sp, and sp², respectively.

C_2H_4

This compound contains a double bond and two single bonds on each C atom.

There are three regions of electron density around each C atom. The double bond is one region.

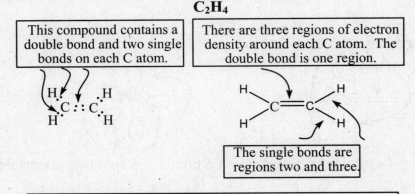

The single bonds are regions two and three.

Just as for BH_3, there are three regions of high electron density surrounding the C atom and the atom is sp² hybridized.

90

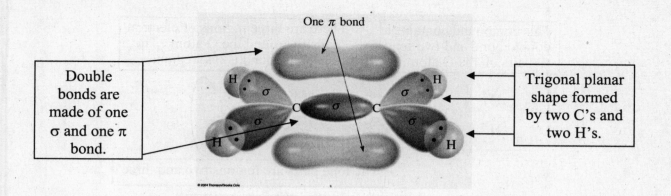

One π bond

Double bonds are made of one σ and one π bond.

Trigonal planar shape formed by two C's and two H's.

C₂H₂

This compound contains a triple bond and one single bond on each C atom.

There are two regions of electron density around each C atom. The triple bond is one region.

H:C:::C:H

H—C≡C—H

The single bond is region two.

Just as for BeH₂, two regions of high electron density surround the C atom and the atom is sp² hybridized.

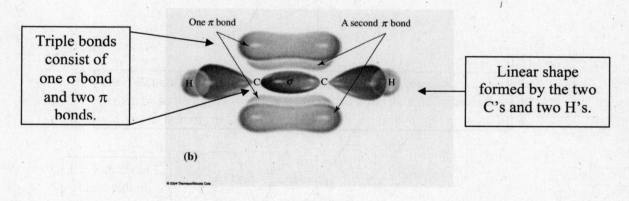

One π bond

A second π bond

Triple bonds consist of one σ bond and two π bonds.

Linear shape formed by the two C's and two H's.

(b)

CH₂O

This compound contains a double bond and two lone pairs on the O atom.	There are three regions of electron density around the O atom. The double bond is one region.

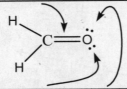

The lone pairs are regions two and three.

Because there are three regions of electron density around the O atom, the hybridization is sp^2.

3. **What is the hybridization of each of the indicated atoms in the amino acid alanine? The correct answer is: atom 1 is sp^3 hybridized, atom 2 is sp^3 hybridized, atom 3 is sp^3 hybridized, atom 4 is sp^2 hybridized, atom 5 is sp^2 hybridized, and atom 6 is sp^3 hybridized**

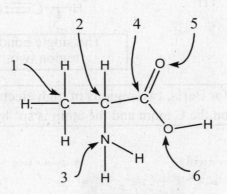

4 single bonds 4 regions of high electron density sp^3 hybrid	2 single bonds & 1 double bond 3 regions of high electron density sp^2 hybrid

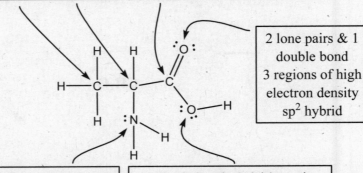

2 lone pairs & 1 double bond 3 regions of high electron density sp^2 hybrid

3 single bonds & 1 lone pair 4 regions of high electron density sp^3 hybrid	2 single bonds & 2 lone pairs 4 regions of high electron density sp^3 hybrid

Polarity of Molecules

4. Which of the following molecules are polar?

BH_3, CH_2F_2, H_2O, SF_6

The correct answer is: BH_3 and SF_6 are nonpolar; CH_2F_2 and H_2O are polar.

YIELD To be polar molecules must have two essential features.
1) The molecule must contain at least one polar bond or one lone pair of electrons.   2) The molecule must be asymmetrical so that the bond dipoles do not cancel each other.

TIP Determining whether or not polar bonds cancel each other can be difficult. Imagine the central atom as a ball with strings attached to it that correlate to the molecule's bonds. If you pull on each string simultaneously, will the ball move? If the answer is yes, then the molecule is polar. Keep in mind that you have to "pull" with different strengths if the atoms attached to the central atom are not all the same.

BH_3 contains polar bonds but the B-H bonds are symmetrical. Thus the dipoles for the polar bonds cancel each other and the molecule is nonpolar.

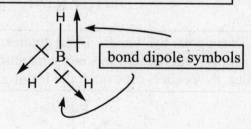

bond dipole symbols

CH$_2$F$_2$ contains 4 polar bonds. The two C-H bonds have their bond dipole pointed toward the C atom (C is more electronegative than H). The two C-F bonds have bond dipoles that are pointed away from the C atom (F is more electronegative than C). The result is an asymmetrical charge distribution making the molecule polar.

INSIGHT: Because CH$_2$F$_2$ is tetrahedral, every possible arrangement of the atoms is polar.

H$_2$O has two polar bonds and two lone pairs. Both bond dipoles for the O-H bonds are directed toward the O atom (O is more electronegative than H). These reinforce the large negative effect of the lone pairs making H$_2$O quite polar.

SF$_6$ has six polar S-F bonds. Bond dipoles are all directed toward the F atoms. But because this is a symmetrical arrangement of the dipoles, they cancel each other out leaving a nonpolar molecule.

Practice Test Three
Modules 7-11

Level 3

1. Determine the electron configuration of Ni, and answer the following questions.

   a) What is the principle quantum number, n, for the d electrons of Ni?
   b) How many of the d electrons are paired?
   c) How many of the d electrons are unpaired?
   d) What is the value of ℓ for the d electrons?

Level 3

2. How many different electrons can have the following combination of the quantum numbers n and ℓ?

   a) $n = 3$, $\ell = 2$
   b) $n = 1$, $\ell = 0$
   c) $n = 5$, $\ell = 3$
   d) $n = 3$, $\ell = 1$
   e) $n = 4$, $\ell = 0$

Levels 1-3

3. Select both the most electronegative element and the element with the greatest first ionization energy from the list below.
   Na, O, N, Al

Level 3

4. Select the element for which the most energy is released when it accepts an electron. Does this element have a very positive or a very negative electron affinity value? Rb, Cs, I, Cl

Level ?

5. Explain why atomic radii increase down a group and from right to left across a period.

Levels 1-2

6. Draw the Lewis structure for MgO. Is MgO covalent or ionic? How many ions are present in one molecule of MgO?

Levels 1-2

7. Draw the Lewis structure for SF_4. Is this molecule covalent or ionic? How many ions are present in one molecule of SF_4?

Levels 1-2

8. Determine both the electronic and the molecular geometries of the following molecules/ions: XeF_4, I_3^-, CO_2, C_2H_2

Level 1

9. Which of the molecules/ions in question 8 are polar?

Level 1

10. Determine the orbital hybridization for the central atom of each molecule in question 8 (there are two central atoms in C_2H_2).

95

Module 12 Predictor Questions

The following questions may help you to determine to what extent you need to study this module. The questions are ranked according to ability.

Level 1 = basic proficiency
Level 2 = mid level proficiency
Level 3 = high proficiency

If you can correctly answer the Level 3 questions, then you probably do not need to spend much time with this module. If you are only able to answer the Level 1 problems, then you should review the topics covered in this module.

Level 1
1. Identify the following as Arrhenius acids, Arrhenius bases, or neither.
   a) H_3BO_3
   b) RbOH
   c) $Ca(OH)_2$
   d) $C_2H_3O_2H$
   e) NaOH
   f) HNO_3

Level 1
2. According to the Brønsted-Lowry theory, a base is defined as:
   a) an electron pair acceptor
   b) a proton acceptor
   c) an electron pair donor
   d) a proton donor
   e) any species that can produce hydroxide ions in aqueous solution

Level 1
3. According to Lewis theory, a base is defined as:
   a) a proton acceptor
   b) a proton donor
   c) an electron-pair donor
   d) any compound that contains electron pairs
   e) an electron-pair acceptor

Level 1
4. Which of the following species *could not* react as a Brønsted-Lowry acid? HCl, H_2O, CaO, NH_4^+, CH_3COOH

Level 2
5. Which one of the following is a weak acid?
   HNO_3, H_3PO_4, $HClO_3$, $HClO_4$, HI

Level 3　　　　6. In the following reaction, CH_3NH_2 can be classified as:
a) only an Arrhenius base
b) only a Lewis acid
c) only a Brønsted-Lowry base
d) a Brønsted-Lowry base and a Lewis base
e) Arrhenius, Brønsted-Lowry, and Lewis bases

Level 3　　　　7. According to the Brønsted-Lowry theory, which of these anions is the strongest base?
NO_3^-, Cl^-, CN^-, ClO_4^-, HSO_4^-

Level 2　　　　8. Identify the conjugate acid-base pairs in the reactions below.
a) $H_2S + NH_3 \rightarrow HS^- + NH_4^+$
b) $H_2O + SO_3^{2-} \rightarrow OH^- + HSO_3^-$
c) $HF + H_2O \rightarrow F^- + H_3O^+$

Level 3　　　　9. Arrange the following in order of increasing base strength.
HSO_4^-, $HSeO_4^-$

Module 12 Predictor Question Solutions

1. Arrhenius acids are molecules that contain hydrogen and dissociate in water to produce H^+. Arrhenius bases contain hydroxyl groups and dissociate in water to produce OH^- ions.

a) H_3BO_3	Arrhenius acid	
b) RbOH	Arrhenius base	
c) $Ca(OH)_2$	Arrhenius base	
d) $C_2H_3O_2H$	Arrhenius acid	
e) NaOH	Arrhenius base	
f) HNO_3	Arrhenius acid	

2. The answer is b). Bronsted-Lowry bases are proton acceptors.

3. The answer is c). Lewis bases are electron pair donors.

4. CaO cannot act as a Bronsted-Lowry acid because it has no proton (H^+) to donate.

5. H_3PO_4 is a weak acid. You should memorize the seven strong acids and remember that all other acids are weak.

6. CH_3NH_2 accepts a proton (H^+) at the nitrogen atom to form the conjugate acid CH_3NH_3, so it is a Bronsted-Lowry base. In accepting the H^+, CH_3NH_2 also donates an electron pair. Thus, it is also a Lewis base.

7. The easiest way to evaluate base strength is to look at the conjugate acids. Remember that weak acids have strong conjugate bases while strong acids have weak conjugate bases.

Base	Conjugate Acid	Acid Strength
NO_3^-	HNO_3	strong
Cl^-	HCl	strong
CN^-	HCN	weak
ClO_4^-	$HClO_4$	strong
HSO_4^-	H_2SO_4	strong

Since HCN is the only weak acid, its conjugate base (CN^-) must be the strongest base in the list.

8. Conjugate acid-base pairs differ only by the presence (acid) or absence (base) of a proton.

        a) H_2S = acid
           HS^- = conjugate base

           NH_3 = base
           NH_4^+ = conjugate acid

        b) H_2O = acid
           OH^- = conjugate base

           SO_3^{2-} = base
           HSO_3^- = conjugate acid

        c) HF = acid
           F^- = conjugate base

           H_2O = base
           H_3O^+ = conjugate acid

9. Again, look at the conjugate acids in order to evaluate base strength.
The conjugate acid of HSO_4^- is H_2SO_4, a strong acid. The conjugate acid of $HSeO_4^-$ is H_2SeO_4, a weak acid. Since H_2SeO_4 is a weak acid, it has the stronger conjugate base.

$HSeO_4^-$ is a stronger base than HSO_4^-.

Module 12
Acids and Bases

Introduction

There are three common theories of acids and bases that are commonly discussed in general chemistry texts: the Arrhenius theory, the Brønsted-Lowry theory, and the Lewis theory. This module will help you understand:

1. the distinctions and the commonalities between the three theories
2. how to distinguish between compounds that act as acids or bases in one theory but not in another

---

Module 12 Key Concepts

1. **The Arrhenius Acid-Base Theory**

    Acid: produces protons (H^+) in aqueous solution

    Base: produces OH^- in aqueous solution

    This is the most restrictive of the three theories since it requires an aqueous solution and compounds that have either an H^+ or an OH^-.

2. **The Brønsted-Lowry Acid-Base Theory**

    Acid: proton donor

    Base: proton acceptor

    This theory is less restrictive. Bases do not have to contain OH^-, and the compounds do not have to be in aqueous solution.

3. **The Lewis Acid-Base Theory**

    Acid: electron pair donor

    Base: electron pair acceptor

    This is the least restrictive of the theories as it does not require the presence of protons, OH^-, or aqueous species.

Sample Exercise
Arrhenius Acid-Base Theory

1. Which of these compounds are Arrhenius acids and which are Arrhenius bases?
HCl, $NaOH$, H_2SO_4, BCl_3, Na_2CO_3, $Ba(OH)_2$, C_2H_4

The correct answer is: HCl and H_2SO_4 are Arrhenius acids; $NaOH$ and $Ba(OH)_2$ are Arrhenius bases. Na_2CO_3, BF_3 and C_2H_4 are neither Arrhenius acids nor bases.

TIP — To identify Arrhenius acids look for compounds that dissociate or ionize in water forming H^+. To identify Arrhenius bases look for compounds that dissociate or ionize in water producing OH^-.

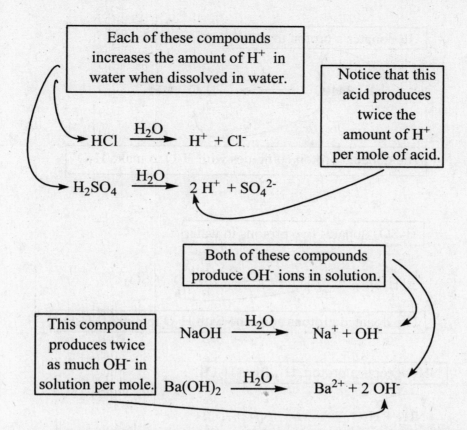

Each of these compounds increases the amount of H^+ in water when dissolved in water.

Notice that this acid produces twice the amount of H^+ per mole of acid.

$$HCl \xrightarrow{H_2O} H^+ + Cl^-$$

$$H_2SO_4 \xrightarrow{H_2O} 2\,H^+ + SO_4^{2-}$$

Both of these compounds produce OH^- ions in solution.

This compound produces twice as much OH^- in solution per mole.

$$NaOH \xrightarrow{H_2O} Na^+ + OH^-$$

$$Ba(OH)_2 \xrightarrow{H_2O} Ba^{2+} + 2\,OH^-$$

CAUTION

It is relatively easy to see that BF_3 and Na_2CO_3 are not acids or bases under this theory since they do not contain H or OH. C_2H_4 may be a little trickier. It does contain H atoms. Do not let this confuse you! The H atoms are not acidic in this case because the C-H bond is too strong to be easily broken.

Brønsted-Lowry Acid-Base Theory
2. *Which of these compounds can be classified as Brønsted-Lowry acids and bases?*
   HF, NH₃, H₂SO₄, BCl₃, Na₂CO₃, K₂S

The correct answer is: HF and H_2SO_4 are Brønsted-Lowry acids; NH_3 and Na_2CO_3 are Bronsted-Lowry bases. BCl_3 and K_2S are neither.

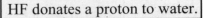

HF donates a proton to water.

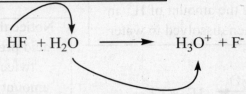

$$HF + H_2O \longrightarrow H_3O^+ + F^-$$

The donated proton combines with H_2O to make H_3O^+.

H_2SO_4 donates two protons to water

$$H_2SO_4 + 2\ H_2O \longrightarrow 2\ H_3O^+ + SO_4^{2-}$$

The donated protons combine with H_2O to make $2\ H_3O^+$.

NH_3 accepts a proton, H^+, from H_2O.

$$NH_3 + H_2O \longrightarrow NH_4^+ + OH^-$$

The proton combines with NH_3 to form NH_4^+.

NH_3 accepts a proton, H^+, from HCl.

$$NH_3 + HCl \longrightarrow NH_4Cl_{(s)}$$

The proton combines with NH_3 to form NH_4^+. This reaction is an example of a <u>nonaqueous</u> Brønsted-Lowry acid-base reaction.

Carbonate ion, CO_3^{2-}, accepts a proton from water.

$$CO_3^{2-} + H_2O \longrightarrow HCO_3^- + OH^-$$

The proton combines with CO_3^{2-} to form HCO_3^-. The Na^+ ions are spectator ions in this reaction.

YIELD
- Because Arrhenius and Brønsted-Lowry acids are both proton donors, they can both be identified similarly.
- Brønsted-Lowry bases, however, may not contain hydroxide ions. Instead they will accept a proton from water, causing hydroxide ions to form in aqueous solutions.

Conjugate acid-base pairs: Under the Brønsted-Lowry theory, these are two species that differ by the presence or absence of a proton. Each Brønsted-Lowry acid has a conjugate base (the base does not have the proton), and each Brønsted-Lowry base has a conjugate acid.

3. Identify the Brønsted-Lowry acid-base conjugate pairs in these reactions.

$$CH_3COOH + H_2O \rightleftarrows CH_3COO^- + H_3O^+$$

$$F^- + H_2O \rightleftarrows HF + OH^-$$

The correct answer is: CH₃COOH is an acid; CH₃COO⁻ is its conjugate base
H₂O is a base; H₃O⁺ is its conjugate acid
F⁻ is a base; HF is its conjugate acid
H₂O is an acid; OH⁻ is its conjugate base

$$CH_3COOH + H_2O \rightleftarrows CH_3COO^- + H_3O^+$$

CH_3COOH donates a proton to H_2O making it an acid. It's conjugate base, CH_3COO^-, differs from the acid by the loss of a single proton, H^+.

$$CH_3COOH + H_2O \quad\rightleftarrows\quad CH_3COO^- + H_3O^+$$

H_2O accepts a proton from CH_3COOH making it a base. It's conjugate acid, H_3O^+, differs from the base by the addition of a single proton, H^+.

INSIGHT: | Many textbooks use this symbolism to designate the acid-base pairs. The 1's indicate that CH_3COOH and CH_3COO^- form one acid-base conjugate pair and the 2's indicate the members of the second acid-base pair.

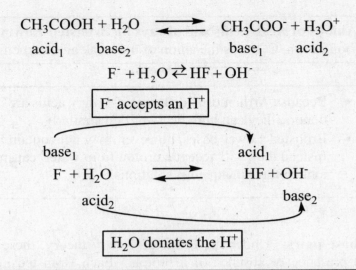

$$\text{CH}_3\text{COOH} + \text{H}_2\text{O} \rightleftharpoons \text{CH}_3\text{COO}^- + \text{H}_3\text{O}^+$$

acid₁   base₂   base₁   acid₂

$$\text{F}^- + \text{H}_2\text{O} \rightleftharpoons \text{HF} + \text{OH}^-$$

F⁻ accepts an H⁺

base₁   acid₁

$$\text{F}^- + \text{H}_2\text{O} \rightleftharpoons \text{HF} + \text{OH}^-$$

acid₂   base₂

H₂O donates the H⁺

YIELD

Notice that in one reaction H_2O is a base and in the other reaction it is an acid. This is called an *amphoteric* species. **In Brønsted-Lowry theory all acid-base reactions are a competition between stronger and weaker acids or bases**. In the CH_3COOH reaction, the stronger acid is CH_3COOH, thus water acts as a base in its presence. In the F^- reaction, H_2O is the stronger acid so it acts as the acid in this reaction. Water is an amphoteric species; it can be either an acid or a base in the presence of a stronger acid or base.

4. *Arrange the following species in order of increasing base strength: HCO_3^-, Cl^-, CO_3^{2-}?*
   The correct answer is: $\text{Cl}^- < \text{HCO}_3^- < \text{CO}_3^{2-}$

INSIGHT:

The best approach to a problem like this is to recognize from which acid or base the species are derived. (You can do this by adding H^+ to the species. In other words, determine the conjugate acids of each species.)

The Cl^- ion is derived from the acid HCl, and HCO_3^- is from H_2CO_3. CO_3^{2-} derives from HCO_3^-.

Now we can easily compare the strengths of the conjugate acids and determine the base strengths.

-For acid strengths, HCl is by far the strongest acid, H_2CO_3 is the next strongest acid, and finally the HCO_3^- ion is the weakest acid. In fact, the HCO_3^- ion is a weak base.

-Since the Cl^- ion is the conjugate base of the strong acid HCl, it is the weakest base. The HCO_3^- ion is the conjugate base of the weak acid H_2CO_3, thus it is a stronger base than Cl^-. Finally, the CO_3^{2-} ion is the conjugate base of the very weak acid (a basic compound is a very weak acid) HCO_3^- making it the strongest base.

Lewis Acid-Base Theory

5. *Identify the Lewis acids and bases in the following reactions.*

$$NH_3 + HCl \rightarrow NH_4Cl$$
$$BCl_3 + NH_3 \rightarrow BCl_3NH_3$$

The correct answer is: HCl and BCl₃ are Lewis acids; NH₃ is the Lewis base in both reactions

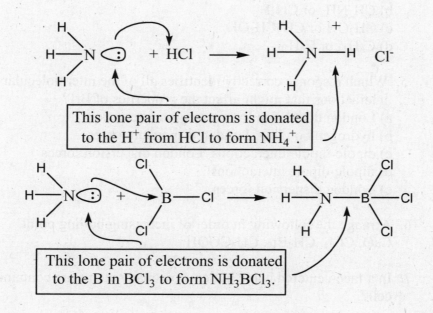

This lone pair of electrons is donated to the H^+ from HCl to form NH_4^+.

This lone pair of electrons is donated to the B in BCl₃ to form NH_3BCl_3.

Module 13 Predictor Questions

The following questions may help you to determine to what extent you need to study this module. The questions are ranked according to ability.
       Level 1 = basic proficiency
       Level 2 = mid level proficiency
       Level 3 = high proficiency

If you can correctly answer the Level 3 questions, then you probably do not need to spend much time with this module. If you are only able to answer the Level 1 problems, then you should review the topics covered in this module.

Level 1   1. A 6.25 L sample of gas exerts a pressure of 1.46 atm at 25°C. What would the pressure of this gas sample be at 25°C if it were compressed to a volume of 5.05 L?

Level 2   2. If 1.47×10^{-3} mol Ar gas occupies a 75.0 mL container at 26.0°C, what is the pressure in atm?

Level 1   3. What is the mass (in grams) of 207 mL of chlorine trifluoride gas at 0.920 atm and 45.0°C?

Level 3   4. Identify the dominant intermolecular forces for each of the following substances. Then, select the substance with the higher boiling point in each pair.
       a) $MgCl_2$ or PCl_3
       b) CH_3NH_2 or CH_3F
       c) CH_3OH or CH_3CH_2OH
       d) C_6H_{14} or C_6H_{12}

Level 3   5. Which response correctly identifies all of the intermolecular interactions that might affect the properties of BrI?
       a) London dispersion forces, ion-ion interaction
       b) hydrogen bonding, London dispersion forces
       c) dipole-dipole interactions, London dispersion forces
       d) dipole-dipole interactions
       e) London dispersion forces

Level 2   6. Arrange the following in order of increasing boiling point.
       CaO, CCl_4, CH_2Br_2, CH_3COOH

Level 1   7. In a face-centered cubic lattice, how many atoms are contained in a unit cell?

Level 2   8. A solid has a density of 5.42 g/cm^3 and crystallizes in a cubic unit cell with an edge length of 4.46 x 10^{-8} cm. If the substance has a molar mass of 144.24 g/mol, how many atoms are in one cell? Identify the type of cubic unit cell.

Level 1   9. Under comparable conditions, how much faster will a sample of He effuse out through a small opening than a sample of Cl$_2$?

Module 13 Predictor Question Solutions

1.

$$P_1 V_1 = P_2 V_2$$

$$(1.46 \text{ atm})(6.25 \text{ L}) = P_2 (5.05 \text{ L})$$

$$P_2 = 1.81 \text{ atm}$$

2.

$$T = 26°C + 273.15 = 299.15 \text{ K}$$

$$V = 75.0 \text{ mL} = 0.0750 \text{ L}$$

$$PV = nRT \Rightarrow P = \frac{nRT}{V} = \frac{(1.47 \times 10^{-3} \text{ mol})(0.0806 \frac{\text{L} \cdot \text{atm}}{\text{mol} \cdot \text{K}})(299.15 \text{ K})}{0.0750 \text{ L}} = 0.481 \text{ atm}$$

3.

$$T = 26°C + 273.15 = 299.15 \text{ K}$$

$$V = 75.0 \text{ mL} = 0.0750 \text{ L}$$

$$PV = nRT \Rightarrow n = \frac{PV}{RT} = \frac{(0.920 \text{ atm})(0.207 \text{ L})}{(0.08205 \frac{\text{L} \cdot \text{atm}}{\text{mol} \cdot \text{K}})(318.15 \text{ K})} = 7.29 \times 10^{-3} \text{ mol ClF}_3$$

$$(7.29 \times 10^{-3} \text{ mol ClF}_3) \left(\frac{92.45 \text{ g ClF}_3}{1 \text{ mol ClF}_3} \right) = 0.674 \text{ g ClF}_3$$

4. a) $MgCl_2$ = ion-ion attraction
   PCl_3 = dipole-dipole attraction

   Ion-ion attraction is the stronger force, so $MgCl_2$ has the greater boiling point.

   b) CH_3NH_2 = hydrogen bonding
   CH_3F = dipole-dipole attraction

   Hydrogen bonding is the stronger force, so CH_3NH_2 has the greater boiling point.

   c) CH_3OH = hydrogen bonding
   CH_3CH_2OH = hydrogen bonding

   Since the intermolecular force is the same, the molecule with the greater molecular weight has the greater boiling point. CH_3CH_2OH has the greater boiling point.

d) hexane = London dispersion forces
   cyclohexane = London disperson forces

Again, since the intermolecular forces are of the same strength, boiling point is determined by molecular weight. Hexane (C_6H_{14}) has a greater molecular weight than cyclohexane (C_6H_{12}), so it has the greater boiling point.

5. The correct answer is c). BrI is a covalent, polar molecule with no hydrogen bonding Ability. Its primary intermolecular force is dipole-dipole interactions. Additionally, all covalent molecules have London dispersion forces.

6. Higher boiling points result from stronger intermolecular forces. The strength of intermolecular forces is as follows: ion-ion attraction > hydrogen bonding > dipole-dipole attractions > London dispersion forces.

   CaO = ion-ion attraction
   CCl_4 = London dispersion forces
   CH_2Br_2 = dipole-dipole attraction
   CH_3COOH = hydrogen bonding

   Boiling points: $CCl_4 < CH_2Br_2 < CH_3COOH < CaO$

7. There are **four** atoms per unit cell in a face centered cubic lattice.

8.
$$V = l^3 = (4.46 \times 10^{-8} \text{ cm})^3 = 8.86 \times 10^{-23} \text{ cm}^3$$

$$D = \frac{m}{V} \Rightarrow m = DV = (5.42 \frac{g}{cm^3})(8.86 \times 10^{-23} \text{ cm}^3) = 4.80 \times 10^{-22} \text{ g}$$

$$\text{mass of one atom} = \left(\frac{144.24 \text{ g}}{1 \text{ mol}}\right)\left(\frac{1 \text{ mol}}{6.022 \times 10^{23} \text{ atoms}}\right) = 2.40 \times 10^{-22} \text{ g/atom}$$

$$\frac{4.80 \times 10^{-22} \text{ g}}{2.40 \times 10^{-22} \frac{g}{atom}} = 2 \text{ atoms}$$

Two atoms indicates a body-centered cubic unit cell.

9.
$$\frac{R_1}{R_2} = \sqrt{\frac{M_1}{M_2}} = \sqrt{\frac{70.90 \text{ g/mol}}{4.00 \text{ g/ mol}}} = 4.21$$

He effuses 4.21 times faster than Cl_2.

Module 13
States of Matter

Introduction

This module describes the basic laws that govern the three states of matter: gas, liquid, and solid. The goals of the module are to:

1. become familiar with how to use the combined and ideal gas laws
2. be able to utilize Graham's law of effusion
3. learn how to determine the relative freezing and boiling points of various liquids based on their intermolecular forces
4. learn to determine the relative melting points of various solids based on the strength of their bonding
5. understand how to use the number of particles in the three cubic unit cells to calculate the radii of atoms.

Module 13 Key Equations & Concepts

1. The combined gas law

$$\frac{P_1 V_1}{T_1} = \frac{P_2 V_2}{T_2}$$

This is a combination of Boyle's and Charles's gas laws. It is used to determine a new temperature, volume, or pressure of a gas given the original temperature, volume and pressure.

2. The ideal gas law

$$PV = nRT$$

This equation is used to calculate the pressure, volume, temperature, or number of moles of a gas given three of the other quantities. It is often used in reaction stoichiometry problems involving gases.

3. Graham's law of effusion

$$\frac{R_1}{R_2} = \sqrt{\frac{M_2}{M_1}}$$

This law is used to determine how quickly one gas effuses (or diffuses) relative to another gas. It can also be used to determine the molar masses of gases based on their effusion rates.

4. Ion-ion interactions, dipole-dipole interactions, hydrogen bonding, London dispersion forces

These are the four basic intermolecular forces involved in liquids. The strength of these interactions determines the boiling points of each liquid.

5. Simple Cubic Unit Cells contain 1 particle per unit cell

The simplest type of cubic unit cell has an atom, ion, or molecule at each of the corners. Because the atoms, ions, and molecules are shared from unit cell to unit cell, each one contributes one-eighth of its volume to a single unit cell. Thus there are 8 x 1/8 =1 atom, ion, or molecule per unit cell.

6. **Body-centered Cubic Unit Cells contain 2 particles per unit cell**

    Body-centered cubic unit cells have one more atom, ion, or molecule in the center of the unit cell. Thus there are (8 x (1/8)) + 1 = 2 atoms, ions, or molecules per unit cell.

7. **Face-centered Cubic Unit Cells contain 4 particles per unit cell**

    Face-centered cubic unit cells have six additional atoms, ions, and molecules (one in each face of the cube.) These atoms, ions, and molecules are shared one-half in each unit cell. Thus there are (8 x (1/8)) + (6 x (1/2)) = 4 atoms, ions, or molecules per unit cell.

8. **Covalent Network Solids, Ionic solids, Metallic solids, Molecular solids**

    These are the four basic types of solids. The strength of the bonds in solids determines the freezing and boiling points of each.

Sample Exercises
Gas Laws

1. *A sample of a gas initially having a pressure of 1.25 atm and volume of 3.50 L has its volume changed to 7.50 x 10⁴ mL at constant temperature. What is the new pressure of the gas sample?*

    The correct answer is: 0.0583 atm.

$$7.50 \times 10^4 \text{ mL} \left(\frac{1 \text{ L}}{1000 \text{ mL}} \right) = 75.0 \text{ L}$$

$$\frac{P_1 V_1}{T_1} = \frac{P_2 V_2}{T_2} \text{ simplifies to } P_1 V_1 = P_2 V_2 \text{ at constant temperature} (T_1 = T_2)$$

$$1.25 \text{ atm} \times 3.50 \text{ L} = P_2 \times 75.0 \text{ L}$$

$$\frac{1.25 \text{ atm} \times 3.50 \text{ L}}{75.0 \text{ L}} = P_2$$

$$0.0583 \text{ atm} = P_2$$

 It is very important that the proper units be used in these problems. In this problem we must change the mL to L or vice versa.

2. *A gas sample initially having a pressure of 1.75 atm and a volume of 4.50 L at a 25.0°C is heated to 37.0°C at a pressure of 1.50 atm. What is the gas's new volume?*
The correct answer is: 5.46 L.

$$\frac{P_1 V_1}{T_1} = \frac{P_2 V_2}{T_2} \text{ where:}$$

$$P_1 = 1.75 \text{ atm}, V_1 = 4.50 \text{ L}, T_1 = 25.0^\circ C = 298.1 \text{ K}$$

$$P_2 = 1.50 \text{ atm and } T_2 = 37.0^\circ C = 310.1 \text{ K}$$

$$V_2 = \frac{P_1 V_1 T_2}{T_1 P_2} = \frac{(1.75 \text{ atm})(4.50 \text{ L})(310.1 \text{ K})}{(298.1 \text{ K})(1.50 \text{ atm})} = 5.46 \text{ L}$$

CAUTION   All gas law problems involving temperature must be in units of Kelvin. Be absolutely certain that you convert temperatures into Kelvin when working any gas law problems.

3. **A gas sample at a pressure of 3.50 atm and a temperature of 45.0°C has a volume of 1.65 x 10³ mL. How many moles of gas are in this sample?**
   The correct answer is: 0.221 moles.

$$PV = nRT \text{ where:}$$

$$P = 3.50 \text{ atm}, V = 1.65 \times 10^3 \text{ mL} = 1.65 \text{ L}, R = 0.0821 \frac{\text{L atm}}{\text{mol K}}, T = 45.0^\circ C = 318.1 \text{ K}$$

$$n = \frac{PV}{RT} = \frac{(3.50 \text{ atm})(1.65 \text{ L})}{\left(0.0821 \frac{\text{L atm}}{\text{mol K}}\right)(318.1 \text{ K})} = 0.221 \text{ mol}$$

INSIGHT: R is the ideal gas constant. In gas laws, its value and units are R = 0.0821 L atm/mol K. This defines the units that we must use in the ideal gas law. P must be in atm, V in L, n in moles, and T in K.

4. **How many grams of $CO_2(g)$ are present in 11.2 L of the gas at STP?**
   The correct answer is: 22.0 g.

INSIGHT: STP is a symbol for standard temperature and pressure. When you see those symbols in a problem involving gases, you may assume that the temperature is 273.15 K and the pressure is 1.00 atm or 760 mm Hg.

$$PV = nRT \text{ thus } n = \frac{PV}{RT}$$

$$n = \frac{(1.00 \text{ atm})(11.2 \text{ L})}{\left(0.0821 \frac{\text{L atm}}{\text{mol K}}\right)(273.15 \text{ K})} = 0.500 \text{ mol}$$

$$0.500 \text{ mol} \left(\frac{44.0 \text{ g CO}_2}{1 \text{ mol CO}_2}\right) = 22.0 \text{ g CO}_2$$

5. If 35.0 g of Al are reacted with excess sulfuric acid, how many L of hydrogen gas, H₂, will be formed at 1.25 atm of pressure and 75.0°C?

$$2 \text{ Al(s)} + 3 \text{ H}_2SO_4(aq) \rightarrow Al_2(SO_4)_3(aq) + 3 \text{ H}_2(g)$$

The correct answer is: 44.6 L.

   a) Calculate the number of moles of hydrogen gas formed in the reaction.

$$(35.0 \text{ g Al})\left(\frac{1 \text{ mol}}{26.98 \text{ g Al}}\right)\left(\frac{3 \text{ mol H}_2}{2 \text{ mol Al}}\right) = 1.95 \text{ mol H}_2$$

   b) Use the ideal gas law to determine the volume of the gas.

$$75.0°C = 273.15 + 75.0°C = 348.1 \text{ K}.$$

$$PV = nRT \text{ thus } V = \frac{nRT}{P}$$

$$V = \frac{(1.95 \text{ mol})\left(0.0821 \frac{\text{L} \cdot \text{atm}}{\text{mol} \cdot \text{K}}\right)(348.1 \text{ K})}{1.25 \text{ atm}} = 44.6 \text{ L}$$

INSIGHT: Sample Exercise 5 uses a combination of reaction stoichiometry and the ideal gas law to determine the volume of the gas formed in a reaction.

6. A gas having a molar mass of 16.0 g/mol effuses through a pinhole 4.00 times faster than an unknown gas. What is the molar mass of the unknown gas?
The correct answer is: 256 g/mol.

Graham's law of effusion relates the rate at which molecules effuse to the molar masses of the substances. In this case R_1 = effusion rate of gas 1, R_2 = effusion rate of gas 2, M_1 = molar mass of gas 1, and M_2 = molar mass of gas 2.

In this problem R_1 = 4.00, and R_2 = 1.00

M_1 = 16.0, and M_2 is unknown.

$$\frac{R_1}{R_2} = \sqrt{\frac{M_2}{M_1}} \text{ thus } \frac{4.00}{1.00} = \sqrt{\frac{M_2}{16.0 \text{ g/mol}}} \text{ to find } M_2 \text{ square both sides of this equation}$$

$$16.0 = \frac{M_2}{16.0 \text{ g/mol}} \text{ thus } M_2 = 16.0 \times 16.0 \text{ g/mol} = 256 \text{ g/mol}$$

INSIGHT:	Deciding which gas has the faster rate can be confusing. However, as long as you associate the rate with one of the gases, i.e. R_1 with M_1 or R_2 with M_2, the relationship will work correctly.

Liquids

7. **Arrange these substances by increasing boiling point: CO_2, NaCl, C_2H_5OH, CH_3Cl**
   The correct answer is: $CO_2 < CH_3Cl < C_2H_5OH < NaCl$

Boiling points are determined by the strength of the intermolecular forces present in a liquid. In general, the strength of intermolecular forces is: ion-ion interactions > hydrogen bonding > dipole-dipole interactions > London dispersion forces.

The primary intermolecular force between molecules of a given substance is determined by the type of compound and its polarity.

Primary Intermolecular Force	Type of Molecule
Ion-ion	Ionic compounds
Hydrogen bonding	Molecules with at least one H directly bonded to O, N, or F atom
Dipole-dipole	Polar molecules
London dispersion forces	Nonpolar molecules

Ion-ion interactions are the strongest of these, and the intermolecular forces get weaker going down the column. This correlates to a decrease in boiling and/or melting points.

The strongest intermolecular forces in liquid CO_2 are London dispersion forces, CH_3Cl's strongest intermolecular forces are dipole-dipole interactions, hydrogen bonding is dominant in C_2H_5OH, and NaCl is an ionic compound. Thus the correct order is: $CO_2 < CH_3Cl < C_2H_5OH < NaCl$.

Solids

8. *Lead, Pb, has a density of 11.35 g/cm³. Solid Pb crystallizes in a unit with an edge length of 4.95 x 10⁻⁸ cm. Which of the three unit cells (simple, body-centered, or face-centered) is present in solid Pb? What is the radius, in cm, of a Pb atom?* The correct answer is: face-centered cubic and the radius is 1.75 x 10⁻⁸ cm.

a) First we need to determine the volume of a single unit cell.

For cubes $V = \ell^3$.

$$V = \left(4.95 \times 10^{-8} \text{ cm}\right)^3 = 1.21 \times 10^{-22} \text{ cm}^3$$

b) Use the volume and the density to determine the mass of a single unit cell.

$$? \text{ g} = 1.21 \times 10^{-22} \text{ cm}^3 \left(\frac{11.35 \text{ g}}{\text{cm}^3}\right) = 1.37 \times 10^{-21} \text{ g}$$

c) Determine the mass of a single Pb atom.

$$? \text{ g} = 207.2 \text{ g/mol} \left(\frac{1 \text{ mol}}{6.022 \times 10^{23} \text{ atoms}}\right) = 3.44 \times 10^{-22} \text{ g/atom}$$

d) Use steps b and c to determine the number of atoms in a single unit cell.

$$? \text{ atoms} = \frac{1.37 \times 10^{-21} \text{ g}}{3.44 \times 10^{-22} \text{ g/atom}} = 3.98 \text{ atoms} \approx 4 \text{ atoms in the unit cell} = \text{face centered cubic unit cell}$$

Type of unit cell	Particles per unit cell
Simple cubic	1
Body-centered cubic	2
Face-centered cubic	4

e) To calculate the radius of a single Pb atom requires use of the Pythagorean theorem and some knowledge of the geometry of a face - centered cubic unit cell.

(ii)

Notice that in this picture of one face of a face-centered unit cell that four atoms form the face diagonal. The face diagonal length is $\sqrt{2}$ times the edge length and that there are 4 atomic radii contained in the face diagonal.

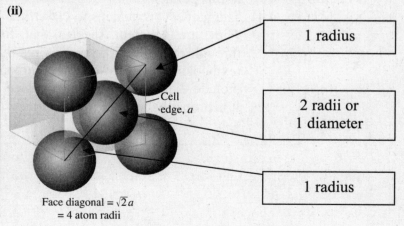

Cell edge, a

1 radius

2 radii or 1 diameter

1 radius

Face diagonal = $\sqrt{2}\,a$
= 4 atom radii

As shown in the diagram : diagonal $= (\sqrt{2})a = 4$ atomi radii

diagonal length $= \sqrt{2}(4.95 \times 10^{-8} \text{ cm}) = 7.00 \times 10^{-8} \text{ cm}$

radius of a Pb atom $= \dfrac{7.00 \times 10^{-8} \text{ cm}}{4 \text{ radii}} = 1.75 \times 10^{-8} \text{ cm}$

INSIGHT:

For the other cubic unit cells the geometrical relationships are:
1) **simple cubic unit cells**
   atomic radius = ½ cell edge length
2) **body-centered cubic unit cells**
   atomic radius = $\sqrt{3}$ x ¼ cell edge length

9. *Arrange these substances by increasing melting point: CO_2, KCl, Na, SiO_2*
   The correct answer is: CO_2 < Na < KCl < SiO_2

INSIGHT:

Melting points of solids are determined by the strength of the forces bonding them together. In general, the weakest forces are intermolecular forces found in molecular solids like CO_2, next weakest are metallic bonds as in Na, ionic bonds are relatively strong like in KCl, and the strongest are the covalent bonds from atom to atom that bond network covalent species like SiO_2.

YIELD

The key to melting point problems is determining a solid's classification.
1) **Molecular solids** are always covalent compounds that form individual molecules. Most of the covalent species that you have learned up to now are molecular solids.
2) **Metallic solids** are by far the easiest to classify. Look for a metallic element.
3) **Ionic solids** are the basic ionic compounds that you have learned up to this point.
4) The hardest substances to classify are the **network covalent species**. They are covalent species that form extremely large molecules through extended arrays of atoms that are covalently bonded. Most textbooks have a list of the common network covalent solids which include diamond, graphite, tungsten carbide (WC), and sand (SiO_2). It is probably best to know memorize these molecules.

Module 14 Predictor Questions

The following questions may help you to determine to what extent you need to study this module. The questions are ranked according to ability.

Level 1 = basic proficiency
Level 2 = mid level proficiency
Level 3 = high proficiency

If you can correctly answer the Level 3 questions, then you probably do not need to spend much time with this module. If you are only able to answer the Level 1 problems, then you should review the topics covered in this module.

Level 1   1. Which of the following compounds are miscible with water?
CH_3OH, CH_3COOH, CCl_4, CH_3NH_2, $HOCH_2CH_2OH$

Level 1   2. Which of the following compounds is miscible in hexane?
CH_3OH, H_2O, CCl_4, C_8H_{18}, $CaBr_2$

Level 1   3. Choose the statements that are correct, given the following information regarding the solubility of NH_4Cl in water:

$$NH_4Cl(s) \rightarrow NH_4Cl(aq) \quad \Delta H_{dissolution} > 0$$

a) Increasing the temperature will increase the solubility of NH_4Cl in water.
b) Increasing the temperature will decrease the solubility of NH_4Cl in water.
c) Increasing the pressure will increase the solubility of NH_4Cl in water.
d) Increasing the pressure will decrease the solubility of NH_4Cl in water.
e) Increasing the pressure will have no effect on the solubility of NH_4Cl in water.

Level 2   4. a) Is CH_3Cl more soluble in CH_3OH or in $CH_3CH_2CH_2CH_2CH_2CH_2OH$?
b) Is $HOCH_2CH_2OH$ more soluble in hexane (C_6H_{14}) or water?

Level 2   5. Determine the %w/w and the $X_{sulfuric\ acid}$ of an aqueous H_2SO_4 solution that is 5.00 m. (Water is the solvent.)

Level 2   6. A 16.0% w/w $C_6H_{12}O_6$ solution has a density of 1.0624 g/mL. What is the concentration of the solution in molarity?

Level 2   7. Sulfur is readily soluble in carbon disulfide, CS_2. The vapor pressure of pure CS_2 is 2.00 atm at 69.1°C. What is the vapor pressure of a solution made by dissolving 32.0 g of S in 380.0 g of CS_2 at 69.1 °C?

Level 1   8. Calculate the freezing point, in °C, of a solution that contains 8.0 g of sucrose (molar mass = 342 g/mol) in 100. g of water. K_f for water is 1.86 °C/m.

Level 1   9. 21.5 mg of a nonelectrolyte is dissolved in sufficient water at a temperature of 5.0°C to make 150. mL of solution. The osmotic pressure of the solution is 0.200 atm. What is the molar mass of the substance?

Module 14 Predictor Question Solutions

1. Polar molecules are miscible primarily in water. The following from the given list are miscible in water: CH_3OH, CH_3COOH, CH_3NH_2, $HOCH_2CH_2OH$

2. Nonpolar molecules are primarily miscible in nonpolar solvents such as hexane. This includes the following from the given list: CCl_4 and C_8H_{18}

3. $\Delta H_{dissolution} > 0$ indicates that the reaction is endothermic (heat is a reactant). Increasing the temperature for an endothermic reaction increases solubility. Changing the pressure has no effect on the solubilities of liquids or solids. Statements a) and e) are true.

4. a) Because CH_3Cl is polar, its solubility is increased in the more polar solvent. Of the two possible solvents both are polar. However, CH_3OH is more polar than the other alcohol because its nonpolar portion (CH_3-) is smaller than the nonpolar portion of the other alcohol ($CH_3CH_2CH_2CH_2CH_2CH_2-$). Consequently, CH_3Cl will dissolve more completely in CH_3OH.

   b) $HOCH_2CH_2OH$ is also polar and consequently more soluble in a polar solvent. H_2O is a quite polar solvent. Hexane is nonpolar. Consequently, more $HOCH_2CH_2OH$ will dissolve in H_2O than in hexane.

5. Molality, m, is defined as moles of solute per kg of solvent. Therefore, 5.00 m can be written as follows:

$$m = \frac{5.00 \text{ mol } H_2SO_4}{1.00 \text{ kg } H_2O}$$

$$(5.00 \text{ mol } H_2SO_4)\left(\frac{98.09 \text{ g } H_2SO_4}{1 \text{ mol } H_2SO_4}\right) = 490. \text{ g } H_2SO_4$$

Mass of solution $= 1000 \text{ g } H_2O + 490.45 \text{ g } H_2SO_4 = 1.49 \times 10^3 \text{ g solution}$

$$\% \text{ w/w} = \frac{\text{mass solute}}{\text{mass solution}} = \frac{490.45 \text{ g } H_2SO_4}{1.49 \times 10^3 \text{ g solution}} = 32.9\% \text{ } H_2SO_4$$

$$(1000 \text{ g } H_2O)\left(\frac{1 \text{ mol } H_2O}{18.02 \text{ g } H_2O}\right) = 55.5 \text{ mol } H_2O$$

$$X_{H_2SO_4} = \frac{\text{mol solute}}{\text{mol solute} + \text{mol solvent}} = \frac{5.00 \text{ mol } H_2SO_4}{5.00 \text{ mol } H_2SO_4 + 55.5 \text{ mol } H_2O} = 0.0826$$

6.

$$16.0\% \text{ w/w } C_6H_{12}O_6 = \left(\frac{16.0 \text{ g } C_6H_{12}O_6}{100. \text{ g solution}}\right)\left(\frac{1.0624 \text{ g solution}}{1.00 \text{ mL}}\right)\left(\frac{1 \text{ mL}}{10^{-3} \text{ L}}\right)\left(\frac{1 \text{ mol } C_6H_{12}O_6}{180.18 \text{ g } C_6H_{12}O_6}\right) = 0.943 \, M$$

7. $P_{\text{solution}} = X_{\text{solvent}}P^0_{\text{solvent}}$

CS_2 is the solvent; S is the solute

$$(32.0 \text{ g S})\left(\frac{1 \text{ mol S}}{32.07 \text{ g S}}\right) = 0.998 \text{ mol S} \qquad (380.0 \text{ g } CS_2)\left(\frac{1 \text{ mol } CS_2}{76.15 \text{ g } CS_2}\right) = 4.99 \text{ mol } CS_2$$

$$X_S = \frac{\text{mol S}}{\text{mol S} + \text{mol } CS_2} = \frac{0.998 \text{ mol}}{0.998 \text{ mol} + 4.99 \text{ mol}} = 0.167$$

$$X_{CS_2} = 1.00 - X_s = 0.833$$

$$P_{\text{solution}} = X_{\text{solvent}}P^0_{\text{solvent}} = (0.833)(2.00 \text{ atm}) = 1.67 \text{ atm}$$

8. Sucrose is a nonelectrolyte, so $i = 1$

$$(8.0 \text{ g sucrose})\left(\frac{1 \text{ mol sucrose}}{342 \text{ g sucrose}}\right) = 2.3 \times 10^{-2} \text{ mol sucrose}$$

$$100. \text{ g } H_2O = 0.100 \text{ kg } H_2O$$

$$m = \frac{2.3 \times 10^{-2} \text{ mol sucrose}}{0.100 \text{ kg } H_2O} = 0.23m$$

$$\Delta T = iK_f m = (1)(1.86°C/m)(0.23 \, m) = 0.44°C$$

The freezing point of pure $H_2O = 0.00°C$

$$0.00°C - 0.44°C = -0.44°C$$

9.

$$\Pi = MRT$$

$$T = 5.0°C + 273.15 = 278.15 \text{ K}$$

$$M = \frac{\Pi}{RT} = \frac{0.200 \text{ atm}}{(0.0821\frac{L \cdot atm}{mol \cdot K})(278.15 \text{ K})} = 0.00876M$$

$$\left(\frac{0.00876 \text{ mol}}{L}\right)(0.150 \text{ L}) = 0.00131 \text{ mol}$$

$$\text{Molar mass} = \frac{0.0215 \text{ g}}{0.00131 \text{ mol}} = 16.4 \text{ g/mol}$$

Module 14
Solutions

Introduction
This module discusses the properties of solutions. The primary goals are to determine how to:

1. use molecular polarity to predict species solubility in various solvents
2. increase the solubility of a given species in a solvent
3. convert from one concentration unit to another
4. use Raoult's law to predict the vapor pressure of a solution
5. determine the freezing and boiling points of solutions
6. calculate the osmotic pressure of a solution

Module 14 Key Equations & Concepts
Like Dissolves Like

This rule is a statement of the common phenomenon that polar molecules are readily soluble in other polar molecules and that nonpolar molecules are readily soluble in other nonpolar molecule. However, polar molecules are fairly insoluble in nonpolar molecules.

Solute solubility is increased when:
1. **the solvent is *heated* in an *endothermic* dissolution**
2. **the solvent is *cooled* in an *exothermic* dissolution**
3. **the pressure of a gas (in a liquid) is increased**

Concentration Units
1. Molarity

$$M = \frac{\text{moles of solute}}{\text{L of solution}}$$

Used in reaction stoichiometry and osmotic pressure problems

2. Molality

$$m = \frac{\text{moles of solute}}{\text{kg of solvent}}$$

Used in freezing point depression and boiling point elevation problems

3. Percent weight by weight

$$\% \text{ w/w} = \frac{\text{mass of one solution component}}{\text{mass of total solution}} (100)$$

Used for concentrated solutions

4. Mole fraction

$$X_A = \frac{\text{moles of component A}}{\text{total moles of solution}}$$

Used in Raoult's Law

> **Raoult's Law**
>
> $P_{solution} = X_{solvent} P^0_{solvent}$
>
> Used to determine the vapor pressure of a solution containing a nonvolatile solute
>
> **Freezing point depression and Boiling point elevation**
>
> $\Delta T_f = iK_f m$ and $\Delta T_b = iK_b m$
>
> These relationships describe how much the freezing or boiling temperatures of a solution will differ from the pure solvent's freezing and boiling points.
>
> **Osmotic pressure of solutions**
>
> $\Pi = MRT$

Sample Exercises
Solubility of a Solute in a Given Solvent
1. Which of the following substances are soluble in water?

$$SiCl_4, NH_3, C_8H_{18}, CaCl_2, CH_3OH, Ca_3(PO_4)_2$$

The correct answer is: only NH_3, $CaCl_2$, and CH_3OH are soluble in water

The "Like Dissolves Like" rule implies that polar species dissolve in polar species and nonpolar species dissolve in nonpolar species. Consequently, nonpolar species do not dissolve in polar species and polar species do not dissolve in nonpolar species. In this example, NH_3 and CH_3OH are both polar covalent compounds so they will dissolve in the highly polar solvent water. $CaCl_2$ is an ionic compound which is water soluble (the solubility rules also apply in these problems). $SiCl_4$ and C_8H_{18} are both nonpolar covalent compounds thus they are insoluble in water. $Ca_3(PO_4)_2$ is an ionic compound that is insoluble in water. Refresh your memory of the solubility rules if necessary.

> Keep in mind these two important questions. 1) Are the covalent compounds in the problem polar or nonpolar? 2) Are the ionic compounds in the problem soluble or insoluble based on the solubility rules? (Remember that strong acids and bases are also water soluble.)

Increasing the Solubility of a Solute in a Given Solvent
2. Given the equation below, which of the following are correct statements?

$$KI(s) \xrightarrow{H_2O} K^+(aq) + I^-(aq) \qquad \Delta H_{dissolution} > 0$$

The correct answer is: only statements a) and f) are true.

a) *Increasing the temperature of the solvent will <u>increase</u> the solubility of the compound in the solvent.*

b) *<u>Decreasing</u> the temperature of the solvent will <u>increase</u> the solubility of the compound in the solvent.*

c) *Changing the temperature of the solvent will <u>not affect</u> the solubility of the compound in the solvent.*

d) *Increasing the pressure of the solute will increase the solubility of the compound in the solvent.*

e) *Increasing the pressure of the solute will decrease the solubility of the compound in the solvent.*

f) *Increasing the pressure of the solute will not affect the solubility of the compound in the solvent.*

There are several important hints in this problem to help you answer it. The $KI_{(s)}$ indicates that this is a solid dissolving in water. Changing the pressure of liquids and solids has no effect on their solubilities. The positive $\Delta H_{dissolution}$ indicates that this is an <u>endothermic</u> process. Heating the solvent for endothermic dissolutions increases the solubility of the solute.

$\Delta H_{dissolution} < 0$ **is exothermic.** $\Delta H_{dissolution} > 0$ **is endothermic.**

3. Given the following dissolution in water equation, which of these changes in conditions are correct statements?

$$O_2(g) \xrightarrow{\text{H}_2\text{O}} O_2(aq) \quad \Delta H_{dissolution} < 0$$

The correct answer is: only conditions b) and d) are correct

a) *Increasing the temperature of the solvent will increase the solubility of the compound in the solvent.*

b) *Decreasing the temperature of the solvent will increase the solubility of the compound in the solvent.*

c) *Changing the temperature of the solvent will not affect the solubility of the compound in the solvent.*

d) *Increasing the pressure of the solute will increase the solubility of the compound in the solvent.*

e) *Increasing the pressure of the solute will decrease the solubility of the compound in the solvent.*

f) *Increasing the pressure of the solute will not affect the solubility of the compound in the solvent.*

The important hints in this problem are 1) $O_{2(g)}$ indicates that this is a gas dissolving in water. Increasing the pressure of gases has a significant effect on their solubilities. In general, increasing the pressure of a gas will increase its solubility in a liquid. 2) The negative $\Delta H_{dissolution}$ indicates that this is an <u>exothermic</u> process. Heating the solvent for exothermic dissolutions decreases the solubility of the solute. Cooling the solvent increases the solubility of the solute in exothermic dissolutions.

INSIGHT:	1) Pay attention to whether the substance being dissolved is a solid, liquid, or gas. That will tell you if the changing pressure condition is applicable. 2) Pay attention to whether or not the dissolution is endo- or exothermic. That is your hint as to heating or cooling the solvent will increase the solubility of the substance. Both of these effects are ramifications of LeChatelier's principle.

Concentration Unit Conversion

4. An aqueous sulfuric acid solution that is 3.75 M has a density of 1.225 g/mL. What is the concentration of this solution in molality (m), percent by mass (% w/w) of H_2SO_4, and mole fraction ($X_{sulfuric\ acid}$) of H_2SO_4?
The correct answers are: 4.38 *m*, 30.0 % w/w, and $X_{sulfuric\ acid}$ = 0.0730

$$3.75\ M\ H_2SO_4 = \frac{3.75\ \text{moles of}\ H_2SO_4}{1.00\ \text{L of solution}}$$

Masses of the solute and solvent must be separated in order to determine the other concentrations. Molarity tells us the volume of the solution, not the mass. The solution's density will help us calculate the mass of this solution. To make the calculation as easy as possible, we can assume that that we have one liter of this 3.75 M solution.

$$1.00\ \text{L of}\ 3.75\ M\ H_2SO_4 = 1000\ \text{mL}\left(\frac{1.225\ \text{g}}{\text{mL}}\right) = 1225\ \text{g of}\ 3.75\ M\ H_2SO_4$$

mass of the solution

$$3.75\ \text{mole}\ H_2SO_4\left(\frac{98.1\ \text{g}\ H_2SO_4}{1\ \text{mole}\ H_2SO_4}\right) = 368\ \text{g}\ H_2SO_4$$

mass of the solute

$$1225\ \text{g} - 368\ \text{g} = 857\ \text{g or}\ 0.857\ \text{kg of water, the solvent}$$

mass of the solvent

$$m = \frac{\text{moles of solute}}{\text{kg of solvent}} = \frac{3.75\ \text{moles of}\ H_2SO_4}{0.857\ \text{kg of}\ H_2O} = \textbf{4.38 m}\ H_2SO_4$$

$$\%\ \text{w/w} = \frac{\text{mass of}\ H_2SO_4}{\text{mass of solution}} \times 100 = \frac{368\ \text{g of}\ H_2SO_4}{1225\ \text{g of solution}} \times 100 = \textbf{30.0\%}\ H_2SO_4$$

$$\text{moles of}\ H_2O = 857\ \text{g}\left(\frac{1\ \text{mole of}\ H_2O}{18\ \text{g of}\ H_2O}\right) = \textbf{47.6 moles of}\ H_2O$$

moles of solvent for the mole fraction

$$X_{sulfuric\ acid} = \frac{\text{moles of solute}}{\text{moles of solute} + \text{moles of solution}} = \frac{3.75\ \text{moles}}{3.75\ \text{moles} + 47.6\ \text{moles}} = \textbf{0.0730}$$

YIELD Converting solution concentrations from molarity to the other three concentration units is by far the hardest type of these conversion problems. The key to doing this correctly is separating the masses of the solute and solvent from the mass of the solution.

5. An aqueous sucrose, $C_{12}H_{22}O_{11}$, solution that is 11.0 % w/w has a density of 1.0432 g/mL. What is the concentration of this solution in molarity (M), molality (m), and mole fraction ($X_{sucrose}$) of $C_{12}H_{22}O_{11}$?
The correct answer is 0.335 *M*, 0.361 *m*, and 0.00646 $X_{sucrose}$

If we assume that we have 100.0 g of solution, we can conclude that we have 11.0 g of sucrose and 89.0 g of water. This is the key to solving this problem.

$$11.0 \text{ g of } C_{12}H_{22}O_{11}\left(\frac{1 \text{ mole of } C_{12}H_{22}O_{11}}{342.3 \text{ g of } C_{12}H_{22}O_{11}}\right) = 0.0321 \text{ moles of } C_{12}H_{22}O_{11}, \text{ the } \textbf{\textit{solute}}$$

$$89.0 \text{ g of } H_2O\left(\frac{1 \text{ mole of } H_2O}{18.0 \text{ g of } H_2O}\right) = 4.94 \text{ moles of } H_2O, \text{ the } \textbf{\textit{solvent}}$$

converting the solution mass to volume

$$\text{volume of } 100.0 \text{ g of this } \textbf{\textit{solution}} = 100.0 \text{ g}\left(\frac{1.00 \text{ mL}}{1.0432 \text{ g}}\right) = 95.9 \text{ mL} = 0.0959 \text{ L}$$

concentrations are easily determined from masses, moles, and volume

$$M = \frac{\text{moles of sucrose}}{\text{L of solution}} = \frac{0.0321 \text{ moles of sucrose}}{0.0959 \text{ L of solution}} = \textbf{0.335 } \textbf{\textit{M}}$$

$$m = \frac{\text{moles of sucrose}}{\text{kg of solvent}} = \frac{0.0321 \text{ moles of sucrose}}{0.0890 \text{ kg of water}} = \textbf{0.361 } \textbf{\textit{m}}$$

$$X_{\text{sucrose}} = \frac{\text{moles of sucrose}}{\text{moles of sucrose + moles of water}} = \frac{0.0321 \text{ moles}}{0.0321 \text{ moles} + 4.94 \text{ moles}} = \textbf{0.00646}$$

INSIGHT: Notice that this problem is much easier because % w/w is a concentration unit that easily separates into the solute and the solvent.

Raoult's Law
6. *What is the vapor pressure of a solution made by dissolving 11.0 g of sucrose in 89.0 g of water at 25.0°C? The vapor pressure of pure water at 25.0°C is 23.76 torr.*
   The correct answer is: 23.60 torr

From exercise 5 above, we know that the $X_{\text{sucrose}} = 0.00646$. Thus, the mole fraction of the solvent, water = $1.00000 - 0.00646 = 0.99354$

$$P_{\text{solution}} = X_{\text{solvent}}P_{\text{solvent}}^{0}$$
$$P_{\text{solution}} = 0.99354\left(23.76 \text{ torr}\right) = 23.60 \text{ torr}$$

mole fraction of the solvent   vapor pressure of the pure solvent

Freezing Point Depression and Boiling Point Elevation
7. *If 11.0 g of sucrose, a nonelectrolyte, are dissolved in 89.0 g of water, at what temperature will this solution boil under 1.00 atm of pressure? The boiling point elevation constant, K_b, for water is 0.512 °C/m.*
   The correct answer is: 100.185 °C.

This solution's concentration was determined in exercise 5 to be 0.361 *m*.

The **van't Hoff factor, *i*,** indicates the extent to which the solute dissociates. For an ionic compound, *i*, is the number of ions in the compound. For nonelectrolytes, *i* is 1.

125

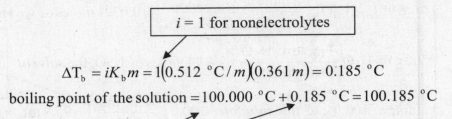

$i = 1$ for nonelectrolytes

$$\Delta T_b = iK_b m = 1(0.512 \ ^\circ C/m)(0.361 \ m) = 0.185 \ ^\circ C$$

boiling point of the solution $= 100.000 \ ^\circ C + 0.185 \ ^\circ C = 100.185 \ ^\circ C$

boiling point of pure water

boiling point increase due to sucrose

8. *12.4 g of a nonelectrolyte are dissolved in 100.0 g of water and the solution is then frozen. The freezing point of the solution is determined to be -5.00 °C. What is the molar mass of the nonelectrolyte? The freezing point depression constant, K_f, for water is 1.86 °C/m.*

The correct answer is: 46.1 g/mol.

Rearranging the relationship to use the quantities given in the problem.

$$\Delta T_f = iK_f m \text{ thus } m = \frac{\Delta T_f}{iK_f} = \frac{\Delta T_f}{K_f} \text{ (for nonelectrolytes)}$$

$$\Delta T_f = 0.00 \ ^\circ C - (-5.00 \ ^\circ C) = 5.00 \ ^\circ C$$

This is the definition of molality.

$$m = \frac{5.00 \ ^\circ C}{1.86 \ ^\circ C/m} = 2.69 \ m$$

$$2.69 \ m = \frac{\text{moles of nonelectrolyte}}{\text{kg of solvent, water}} = \frac{? \text{ moles of nonelectrolyte}}{0.100 \text{ kg of water}}$$

$? \text{ moles of nonelectrolyte} = 2.69 \ m \ (0.100 \text{ kg of water}) = 0.269 \text{ moles of nonelectrolyte}$

$$\text{molar mass of nonelectrolyte} = \frac{\text{mass of nonelectrolyte}}{\text{moles of nonelectrolyte}} = \frac{12.4 \text{ g}}{0.269 \text{ mol}} = 46.1 \text{ g/mol}$$

9. *A 1.00 m solution of a strong electrolyte dissolved in 100.0 g of water forms a solution having a freezing point of -5.58°C. Which of these generic ionic formulas would correspond to the formula of the electrolyte? (M represents a typical metal cation and X a typical anion. The freezing point depression constant, K_f, for water is 1.86 °C/m.)*

The correct answer is: b) MX₂

       a) MX
       b) MX₂
       c) MX₃
       d) M₂X₃
       e) M₂X₄

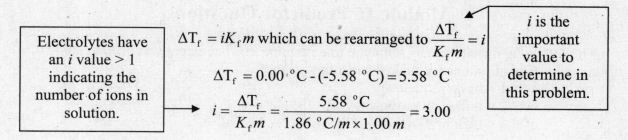

| Electrolytes have an i value > 1 indicating the number of ions in solution. | $\Delta T_f = iK_f m$ which can be rearranged to $\dfrac{\Delta T_f}{K_f m} = i$ $\Delta T_f = 0.00\ ^\circ C - (-5.58\ ^\circ C) = 5.58\ ^\circ C$ $i = \dfrac{\Delta T_f}{K_f m} = \dfrac{5.58\ ^\circ C}{1.86\ ^\circ C/m \times 1.00\ m} = 3.00$ | i is the important value to determine in this problem. |

The answer indicates that this electrolyte has 3 times the effect of a nonelectrolyte on the freezing point depression. Consequently, there must be 3 ions dissolved in solution. In the possible answers only MX_2 can dissolve to generate 3 ions in solution, namely 1 M ion and 2 X ions.

Osmotic Pressure

10. What is the osmotic pressure of a 0.335 M sucrose solution at 25.0°C?
   The correct answer is: 8.20 atm.

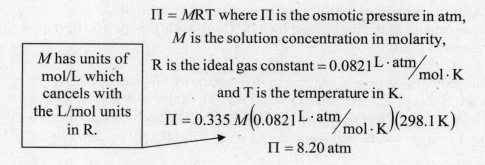

$\Pi = MRT$ where Π is the osmotic pressure in atm,

M is the solution concentration in molarity,

| M has units of mol/L which cancels with the L/mol units in R. | R is the ideal gas constant $= 0.0821\ \mathrm{L \cdot atm}\big/\mathrm{mol \cdot K}$ and T is the temperature in K. $\Pi = 0.335\ M\left(0.0821\ \mathrm{L \cdot atm}\big/\mathrm{mol \cdot K}\right)(298.1\ K)$ $\Pi = 8.20\ \text{atm}$ |

Module 15 Predictor Questions

The following questions may help you to determine to what extent you need to study this module. The questions are ranked according to ability.

      Level 1 = basic proficiency
      Level 2 = mid level proficiency
      Level 3 = high proficiency

If you can correctly answer the Level 3 questions, then you probably do not need to spend much time with this module. If you are only able to answer the Level 1 problems, then you should review the topics covered in this module.

Level 1   1. A chemical reaction releases 58,500 J of heat into 150 g of water. Assuming no heat is lost to the surroundings, what is the temperature *increase* of the water. The specific heat of liquid water is 4.184 J/g.°C.

Level 1   2. A 25.0 g sample of In is heated by exposure to 1.50×10^3 J. The temperature of the In is raised by 258 °C. What is the specific heat of the In in J/g.°C?

Level 1   3. Calculate the amount of heat required to convert 10.0 grams of ice at -20.0°C to 120.°C. The specific heats are: $H_2O(s)$ = 2.09 J/g.°C; $H_2O(\ell)$ = 4.18 J/g.°C; $H_2O(g)$ = 2.03 J/g.°C. The heats of fusion and vaporization are, respectively: $H_2O(s)$ = 333 J/g; $H_2O(\ell)$ = 2260 J/g.

Level 1   4. What is the change in internal energy of a system, in J, if the system emits 763 J of work to its surroundings while absorbing 763 J of heat?

Level 2   5. For one mole of reactions, how much work (in J) is done by the following chemical reaction at constant pressure and a temperature of 32.0°C? Is the work done *on the system* or *by the system?* If ΔH_{rxn} for the reaction is -2219.8 kJ, what are ΔE and q for this system?
          $C_3H_8(g) + 5O_2(g) \rightarrow 3CO_2(g) + 4H_2O(g)$

Level 1   6. Using the table of thermodynamic data provided, calculate the ΔH^0_{rxn} for the following reaction. Determine whether the reaction is endothermic or exothermic.
          $2C_6H_6(\ell) + 15O_2(g) \rightarrow 12 CO_2(g) + 6 H_2O(g)$

Species	ΔH^0_f (kJ/mol)
C_6H_6	49.04
CO_2	-393.5
H_2O	-241.8

128

Level 1

7. Determine ΔS^0_{rxn}, in J/mol, for the combustion of 1 mol of $C_3H_{8\,(g)}$ at 25°C.

$$C_3H_8(g) + 5O_2(g) \rightarrow 3CO_2(g) + 4H_2O(g)$$

Species	ΔS^0_f (J/mol·K)
C_3H_8	269.9
CO_2	213.7
H_2O	69.9
O_2	205.0

Level 1

8. For a certain process at 127°C, ΔG = -16.20 kJ and ΔH = -17.0 kJ. What is the entropy change, in J/K, for this process at this temperature?

Level 1

9. Using the data given below, determine the ΔG^0_{rxn} for this chemical reaction. Determine whether the reaction is spontaneous or nonspontaneous.

$$CS_2(g) + 3O_2(g) \rightarrow CO_2(g) + 2SO_2(g)$$

Species	ΔG^0_f (kJ/mol)
CS_2	67.15
O_2	0.0
CO_2	-394.4
SO_2	-300.2

Module 15 Predictor Question Solutions

1. $q = mC\Delta T \Rightarrow \Delta T = \dfrac{q}{mC} = \dfrac{58500\ J}{(150.\ g)(4.184\dfrac{J}{g \cdot ^\circ C})} = 93.2\ ^\circ C$

2. $q = mC\Delta T \Rightarrow C = \dfrac{q}{m\Delta T} = \dfrac{1.50 \times 10^3\ J}{(25.0\ g)(258\ ^\circ C)} = 0.233\dfrac{J}{g \cdot ^\circ C}$

3. Five steps are required: a) heat ice from -20.0°C to 0.00 °C; b) melt ice; c) heat water from 0.00 °C to 100.0 °C; d) evaporate water; e) heat steam from 100.0 °C to 120.0 °C.

   a) $q = mC\Delta T = (10.0\ g)(2.09\dfrac{J}{g \cdot ^\circ C})(20.0\ ^\circ C) = 418\ J$

   b) $q = m\Delta H_{fusion} = (10.0\ g)(333\dfrac{J}{g}) = 3330\ J$

   c) $q = mC\Delta T = (10.0\ g)(4.18\dfrac{J}{g \cdot ^\circ C})(100.\ ^\circ C) = 4180\ J$

   d) $q = m\Delta H_{evaporation} = (10.0\ g)(2260\dfrac{J}{g}) = 22600\ J$

   e) $q = mC\Delta T = (10.0\ g)(2.03\dfrac{J}{g \cdot ^\circ C})(10.0\ ^\circ C) = 406\ J$

   Sum of five steps $= 3.09 \times 10^4\ J = 30.9\ kJ$

4. $E = q + w = 763\ J + (-763\ J) = 0$

5.

   $\Delta n = \Sigma mol\ gas\ (product) - \Sigma mol\ gas\ (reactant) = 7\ mol - 6\ mol = 1\ mol$

   $T = 32.0\ ^\circ C + 273.15 = 305.15\ K$

   $w = -\Delta n_{gas}RT = -(1\ mol)(8.314\dfrac{J}{mol \cdot K})(305.15\ K) = -2.54 \times 10^3\ J$

   The negative sign indicates that work is done *by* the system.

   $\Delta H_{rxn} = q\ (at\ constant\ pressure) = -2219.8\ kJ)$
   $\Delta E = w + q = -2.54 \times 10^3\ J + -2.22 \times 10^6\ J = -2.22 \times 10^6\ J$

6.

   $\Delta H^0_{rxn} = \Sigma nH^0_{f\ products} - \Sigma nH^0_{f\ reactants}$
   $\Delta H^0_{rxn} = [12(-395.5\ kJ/mol) + 6(-241.8\ kJ/mol)] - [2(49.04\ kJ/mol)] = -6.271 \times 10^3\ kJ$

7.

$$\Delta S^0_{\text{rxn}} = \Sigma n S^0_{\text{f products}} - \Sigma n S^0_{\text{f reactants}}$$

$$\Delta S^0_{\text{rxn}} = [3(213.7 \text{ J/mol} \cdot \text{K}) + 4(69.9 \text{ J/mol} \cdot \text{K})] - [5(205.0 \text{ J/mol} \cdot \text{K}) + 1(269.9 \text{ J/mol} \cdot \text{K})]$$

$$\Delta S^0_{\text{rxn}} = -374.2 \text{ J/K}$$

8.

$$\Delta G = \Delta H - T\Delta S \Rightarrow \Delta S = \frac{-\Delta G + \Delta H}{T}$$

$$T = 127°C + 273.15 = 400.15 \text{ K}$$

$$\Delta S = \frac{-(-16.20 \text{ kJ}) + (-17.0 \text{ kJ})}{400.15 \text{ K}} = -2.00 \times 10^{-3} \text{ kJ/K}$$

9.

$$\Delta G^0_{\text{rxn}} = \Sigma n \Delta G^0_{\text{f products}} - \Sigma n \Delta G^0_{\text{f reactants}}$$

$$\Delta G^0_{\text{rxn}} = [1(-394.4 \text{ kJ/mol}) + 2(-300.2 \text{ kJ/mol})] - [1(67.15 \text{ kJ/mol}) + 0] = -1.062 \times 10^3 \text{ kJ}$$

The negative value for ΔG^0_{rxn} indicates a spontaneous reaction.

Module 15
Heat Transfer, Calorimetry, and Thermodynamics

Introduction

This module presents a brief discussion of heat related topics in chemistry. The following major issues must be addressed. They all focus on how heat and energy are transferred from one chemical system to another.

1. the basic heat transfer equation and its impact on both calorimetry and heating substances that remain in a single phase
2. simple chemical thermodynamics including the change in energy (ΔE) of a system
3. the heat (q) and work (w) involved in an energy change
4. the change in enthalpy (ΔH)
5. calculation of ΔH using Hess's law
6. calculation of the changes in entropy (ΔS) and Gibbs Free Energy (ΔG)
7. the temperature dependence of the Gibbs Free Energy change

---

Module 15 Key Equations & Concepts

1. **$q = mC\Delta T$**
   This is the basic heat transfer equation which calculates the amount of energy emitted or absorbed when an object warms up or cools down. (q is the heat, m is the mass, C is the specific heat, and ΔT is the temperature change.) It is used in calorimetry, heat lost = heat gained problems, and to determine the heat necessary to either heat up or cool down a substance that remains *in a single phase*.

2. **$\Delta E = q + w$**
   The change in the energy of a chemical system is determined by two factors, 1) how much heat (q) enters or leaves the system and 2) how much work (w) the system does in the form of expanding or contracting against a constant pressure such as the atmosphere. The correct signs (i.e., +q or –q) of the heat and work are crucial to understanding these problems.

3. **$w = -P\Delta V = -\Delta n_{gas}RT$ (at constant temperature and pressure)**
   This relationship defines the amount of work that a system can do at constant temperature and pressure. (w is the work, P is the pressure, ΔV is the volume change, Δn_{gas} is the change in the number of moles of gas, R is the ideal gas constant, and T is the temperature.) It also describes the work that a system can do or have done on it when there is a change in the number of moles of gas.

4. **$\Delta H = \Delta E + P\Delta V = q_P$ at constant temperature and pressure**
   This is the definition of enthalpy, ΔH, and its relationship to the energy change of a system. (q_p is the heat flow at constant pressure.)

5. $\Delta H^0_{rxn} = \sum n\Delta H^0_{f\ products} - \sum n\Delta H^0_{f\ reactants}$

This is one form of Hess's law which is used to determine the heat absorbed or released in a chemical reaction from the heats of formation of the products and reactants. ($\Delta H^0_{f\ products}$ is the heat of formation of the product substances at standard conditions. $\Delta H^0_{f\ reactants}$ is the heat of formation of the reactant substances at standard conditions. n represents the stoichiometric coefficients in the balanced chemical reaction.)

6. $\Delta S^0_{rxn} = \sum nS^0_{f\ products} - \sum nS^0_{f\ reactants}$

This is the relationship for determining the entropy change of a chemical reaction given the standard entropies of formation for the products and the reactants under standard conditions.

7. $\Delta G^0_{rxn} = \sum nG^0_{f\ products} - \sum nG^0_{f\ reactants}$

This is the relationship for determining the Gibbs free energy change of a chemical reaction given the standard free energies of formation for the products and the reactants under standard conditions.

8. $\Delta G = \Delta H - T\Delta S$

This is the definition of the Gibbs free energy. It is used to determine the temperature dependence of the free energy.

Sample Exercises
Heat Transfer Calculation

1. *How much heat is required to heat 75.0 g of aluminum, Al, from 25.0°C to 175.0°C? The specific heat of Al is 0.900 J/g °C.*
   The correct answer is: 1.01 x 10⁴ J or 10.1 kJ.

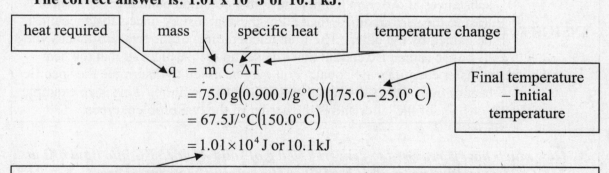

heat required	mass	specific heat	temperature change

$$q = m\ C\ \Delta T$$
$$= 75.0\,g(0.900\,J/g°C)(175.0 - 25.0°C)$$
$$= 67.5J/°C(150.0°C)$$
$$= 1.01 \times 10^4\,J \text{ or } 10.1\,kJ$$

Final temperature – Initial temperature

The heat required is positive indicating that the Al **absorbs** the heat.

CAUTION   When calculating ΔT use $T_{final} - T_{initial}$. This will insure that the sign of q is correct.

133

2. *A 75.0 g piece of aluminum, Al, initially at a temperature of 175.0°C is dropped into a coffee cup calorimeter containing 150.0 g of H₂O initially at a temperature of 15.0°C. What will the final temperature of the system be when it reaches thermal equilibrium? Assume that no heat is lost to the container. The specific heat of Al is 0.900 J/g °C and for water is 4.18 J/g°C.*
The correct answer is: 30.6°C.

$$\text{heat lost by the Al} = \text{heat gained by the H}_2\text{O}$$

$$m_{Al}C_{Al}\Delta T_{Al} = m_{H_2O}C_{H_2O}\Delta T_{H_2O}$$

$$(75.0\,\text{g})(0.900\,\text{J/g}°\text{C})(175.0°\text{C} - T_{final}) = (150.0\,\text{g})(4.18\,\text{J/g}°\text{C})(T_{final} - 15.0°\text{C})$$

$$(11,812.5 - 67.5\,\text{T})\,\text{J} = (627.0\text{T} - 9405.0)\,\text{J}$$

$$11,812.5 + 9405.0 = (627.0\text{T} + 67.5\text{T})$$

$$21,217.5 = 694.5\text{T}$$

$$\frac{21,275}{694.5} = T$$

$$30.6°\text{C} = T$$

The units of Joules cancel.

T_f of the system will be lower than T_i for Al and higher than T_i for H₂O. Set up the problem so that you have the higher T minus the lower T on each side of the equation.

INSIGHT: Heat lost = heat gained problems are characterized by the mixing of two substances at different temperatures in a common container. If no heat is lost to the surroundings, then all of the heat lost by once substance must be gained by the other. The final temperatures of the two substances will be the same. Be certain that you set up the problem so that the heat transfer equations are equal. You may be asked to determine the specific heat of one of the substances or the final temperature. This is an example of the latter and is the harder of the two problem types.

3. *How much heat is required to convert 150.0 g of solid Al at 458°C into liquid Al at 758°C? The melting point of Al is 658°C. The specific heats for Al are, C_{solid} = 24.3 J/mol °C and C_{liquid} = 29.3 J/mol °C. The ΔH_{fusion} for Al = 10.6 kJ/mol.*
The correct answer is: q = 102.2 kJ.

INSIGHT: This problem involves three separate calculation steps. The final answer is the sum of these three steps. The steps are: 1) heat required to warm the Al from 458°C to its melting point, 658°C, 2) heat required to melt the Al, and 3) heat required to heat the liquid Al from its melting point to the final temperature of 758°C. These steps are illustrated in the diagram below.

CAUTION

Remember, phase changes do NOT have temperature changes associated with them!

Heating Curve for Al

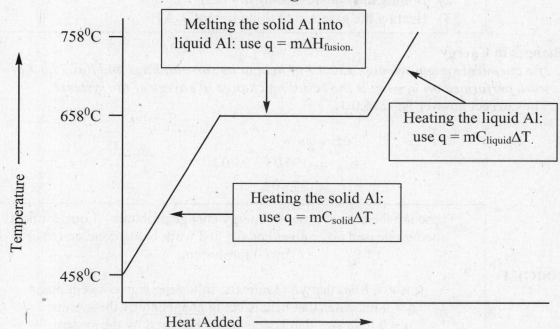

$$150.0 \text{ g}\left(\frac{1 \text{ mol Al}}{26.98 \text{ g}}\right) = 5.560 \text{ mol Al}$$

Because the ΔH and C's are in J/mol, we must convert g Al to mol.

A very common mistake is to not use the correct specific heats for each step!

Step 1) $q = mC_{solid}\Delta T = 5.560 \text{ mol} (24.3 \text{ J/mol } ^\circ C)(658^\circ C - 458^\circ C)$

$= 135 \text{ J/ } ^\circ C (200^\circ C) = \textbf{27.0 kJ}$

Step 2) $q = m\Delta H_{fusion}$

$= 5.560 \text{ mol}(10.6 \text{ kJ/mol}) = \textbf{58.9 kJ}$

Step 3) $q = mC_{liquid}\Delta T = 5.560 \text{ mol}(29.3 \text{ J/mol } ^\circ C)(758^\circ C - 658^\circ C)$

$= 163 \text{ J/}^\circ C (100^\circ C) = \textbf{16.3 kJ}$

Total amount of heat = Step 1 + Step 2 + Step 3 = 27.0 kJ + 58.9 kJ + 16.3 kJ = **102.2 kJ**

Use the correct T range for each heating step!

Sample exercise 3 involves only one phase change, namely converting solid Al into liquid Al. If a second phase change were included, converting solid Al into gaseous Al, the following steps would have to be included.
1) Heating the liquid Al to the boiling point using $q = mC_{liquid}\Delta T$.
2) Boiling the liquid Al using $q = m\Delta H_{vaporization}$.
3) Heating the gaseous Al using $q = mC_{gas}\Delta T$.

Changes in Energy

4. If a chemical system releases 350.0 J of heat to its surroundings and has 75.0 J of work performed on it, what is the resulting change in energy of the system?
The correct answer is: -275.0 J.

$$\Delta E = q + w$$
$$= -350.0 \text{ J} + 75.0 \text{ J}$$
$$= -275.0 \text{ J}$$

INSIGHT: These are the easiest types of energy change problems. Look for heat being released (-) or absorbed (+) and work being done on (+) or by (-) the system.

It is *essential* that you know the following sign conventions
$q > 0$ indicates that heat is being **absorbed** by the system
$q < 0$ indicates that heat is being **released** by the system
$w > 0$ indicates that work is being done **on** the system
$w < 0$ indicates that work is being done **by** the system

Work Involved in a Chemical Reaction

5. How much work is done on or by the system in the following chemical reaction at constant pressure and a temperature of 25.0°C?
$$C_2H_5OH(\ell) \ + \ 3 \ O_2(g) \rightarrow 2 \ CO_2(g) \ + \ 3 \ H_2O(g)$$

The correct answer is: -4.957 x 10³ J; work is being done _by_ the system

INSIGHT: Look for a question that displays a chemical reaction involving a change in the number of moles of gas and asks for the amount of work performed. Be sure that you determine the change in the number of moles of gas as follows:
$$\Delta n_{gas} = \Sigma \text{moles of gas}_{products} - \Sigma \text{moles of gas}_{reactants}$$

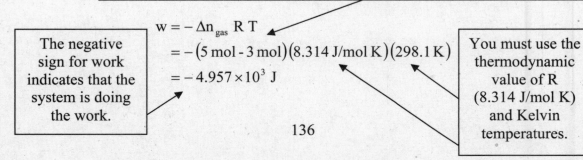

The negative sign for work indicates that the system is doing the work.

$$w = -\Delta n_{gas} R T$$
$$= -(5 \text{ mol} - 3 \text{ mol})(8.314 \text{ J/mol K})(298.1 \text{ K})$$
$$= -4.957 \times 10^3 \text{ J}$$

You must use the thermodynamic value of R (8.314 J/mol K) and Kelvin temperatures.

Relationship of Enthalpy Change to Energy Change

6. *Given the following information about this chemical reaction at constant pressure and a temperature of 25.0°C, what are the values of ΔE, q, and w for this reaction?*

$$C_2H_5OH(\ell) \; + \; 3\,O_2(g) \rightarrow 2\,CO_2(g) \; + \; 3\,H_2O(g) \quad \Delta H^0_{rxn} = -1234.7 \text{ kJ/mol rxn}$$

The correct answer is: q = -1234.7 kJ, w = -4.957 kJ, and ΔE = -1239.3 kJ.

Exercise 5 tells us that the value of w for this reaction is -4.957×10^3 J or -4.957 kJ. The definition of enthalpy change, $\Delta H = \Delta E + P\Delta V = q_P$ gives us a method to determine q.

The $\Delta H^0_{rxn} = q$ at constant pressure = -1234.7 kJ/mol rxn.

$$\Delta E = q + w$$
$$= (-1234.7 \text{ kJ}) + (-4.957 \text{ kJ})$$
$$= -1239.7 \text{ kJ}$$

INSIGHT:	Notice that ΔH and ΔE are almost the same value. They differ only by the amount of work that the system does, -4.957 kJ. This agrees with the definition of enthalpy change, $\Delta H = \Delta E + P\Delta V = \Delta E + \Delta n_{gas}RT$ at constant temperature and pressure.

Calculation of Enthalpy Change for a Reaction

7. *What is the enthalpy change for this reaction at standard conditions?*

$$C_2H_5OH(\ell) \; + \; 3\,O_2(g) \rightarrow 2\,CO_2(g) \; + \; 3\,H_2O(g)$$

The correct answer is: $\Delta H^0_{rxn} = -1234.7$ kJ/mol rxn

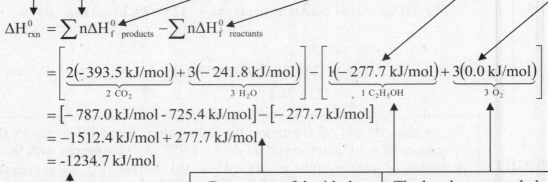

The superscript 0's indicate that these values were measured at standard thermodynamic conditions (1.00 atm of pressure and 273.15 K). The n's are the stoichiometric coefficients from the balanced reaction.

Notice that we must sum the products' ΔH values and the reactants' ΔH values then subtract them. The subscript f's indicate that the ΔH values are for the formation of the substances from their elements.

The ΔH^0_f values are tabulated in an appendix at the back of your textbook. Elements, like O_2, always have a $\Delta H^0_f = 0.0$ kJ/mol.

$$\Delta H^0_{rxn} = \sum n\Delta H^0_{f\ products} - \sum n\Delta H^0_{f\ reactants}$$

$$= \left[\underbrace{2(-393.5\ \text{kJ/mol})}_{2\ CO_2} + \underbrace{3(-241.8\ \text{kJ/mol})}_{3\ H_2O} \right] - \left[\underbrace{1(-277.7\ \text{kJ/mol})}_{1\ C_2H_5OH} + \underbrace{3(0.0\ \text{kJ/mol})}_{3\ O_2} \right]$$

$$= \left[-787.0\ \text{kJ/mol} - 725.4\ \text{kJ/mol} \right] - \left[-277.7\ \text{kJ/mol} \right]$$

$$= -1512.4\ \text{kJ/mol} + 277.7\ \text{kJ/mol}$$

$$= -1234.7\ \text{kJ/mol}$$

The negative value for ΔH^0_{rxn} indicates that this is an **exothermic** reaction.

Be very careful with the signs of the ΔH^0_f values and how they are added and subtracted.

The brackets are to help you understand how the ΔH^0_f values and stoichiometric coefficients are determined.

Calculation of Entropy Change for a Reaction

8. *What is the entropy change for this reaction at standard conditions?*

$$C_2H_5OH(\ell)\ +\ 3\ O_2(g) \rightarrow 2\ CO_2(g)\ +\ 3\ H_2O(g)$$

The correct answer is: $\Delta S^0_{rxn} = 217.3$ J/mol

The n's, superscript 0's, and subscript f's have the same meaning in this equation as in exercise 6. The ΔS^0_f values are also tabulated in an appendix in your text.

$$\Delta S^0_{rxn} = \sum n\Delta S^0_{f\ products} - \sum n\Delta S^0_{f\ reactants}$$

$$= \left[\underbrace{2(213.6\ \text{J/mol K})}_{2\ CO_2} + \underbrace{3(188.7\ \text{J/mol K})}_{3\ H_2O} \right] - \left[\underbrace{1(161.0\ \text{J/mol K})}_{1\ C_2H_5OH} + \underbrace{3(205.0\ \text{J/mol K})}_{3\ O_2} \right]$$

$$= \left[427.2\ \text{J/mol K} + 566.1\ \text{J/mol K} \right] - \left[161.0\ \text{J/mol K} + 615.0\ \text{J/mol K} \right]$$

$$= 993.3\ \text{J/mol K} + 776.0\ \text{J/mol K}$$

$$= 217.3\ \text{J/mol K}$$

The positive value for ΔS^0_{rxn} indicates that this chemical system is more disordered after the reaction has occurred.

Unlike ΔH^0_f, elements can have nonzero values of ΔS^0_f.

Calculation of Gibbs Free Energy Change for a Reaction
9. *What is the Gibbs Free Energy change for this reaction at standard conditions?*
$$C_2H_5OH(\ell) \ + \ 3\ O_2(g) \ \rightarrow \ 2\ CO_2(g) \ + \ 3\ H_2O(g)$$

The correct answer is: ΔG^0_{rxn} = -1299.7 kJ/mol

The ΔG^0_f values are also tabulated in an appendix in your textbook.

$$\Delta G^0_{rxn} = \sum n\Delta G^0_{f\ products} - \sum n\Delta G^0_{f\ reactants}$$

$$= \left[\underbrace{2(-394.4\ kJ/mol)}_{2\ CO_2} + \underbrace{3(-228.6\ kJ/mol)}_{3\ H_2O} \right] - \left[\underbrace{1(-174.9\ kJ/mol)}_{1\ C_2H_5OH} + \underbrace{3(0.0\ kJ/mol)}_{3\ O_2} \right]$$

$$= \left[-788.8\ kJ/mol\ - 685.8\ kJ/mol \right] - \left[-174.9\ J/mol + 0.0\ kJ/mol \right]$$

$$= -1474.6\ kJ/mol + 174.9\ kJ/mol$$

$$= -1299.7\ kJ/mol$$

The negative sign for ΔG^0_{rxn} indicates that this reaction is spontaneous.

Elements have zero values of ΔG^0_f.

It is very important that you know the following sign conventions for ΔH, ΔS, and ΔG.

$\Delta H > 0$ indicates that the process is **endothermic**.
$\Delta H < 0$ indicates that the process is **exothermic**.
$\Delta S > 0$ indicates that the process is **less ordered**.
$\Delta S < 0$ indicates that the process is **more ordered**.
$\Delta G > 0$ indicates that the process is **nonspontaneous**.
$\Delta G < 0$ indicates that the process is **spontaneous**.

YIELD

TIP

Values for ΔH°_f, ΔS°_f, and ΔG°_f can be found in your textbook appendices. Of the three, only ΔS°_f for a substance in its elemental state may have a non-zero value.

Temperature Dependence of Spontaneity
10. *Can this reaction become nonspontaneous if the temperature is changed?*
$$C_2H_5OH(\ell) \ + \ 3\ O_2(g) \ \rightarrow \ 2\ CO_2(g) \ + \ 3\ H_2O(g)$$

The correct answer is no.

From exercises 6 and 7 we see that for this reaction $\Delta H^0_{rxn} < 0$ and $\Delta S^0_{rxn} > 0$. The definition of ΔG^0_{rxn} is $\Delta G^0_{rxn} = \Delta H^0_{rxn} - T\Delta S^0_{rxn}$. In this case, ΔG^0_{rxn} = (negative quantity) – T (positive value) which must always give $\Delta G^0_{rxn} < 0$. (Remember, $\Delta G^0_{rxn} < 0$ indicates a spontaneous reaction.)

YIELD

The following conclusions can be drawn regarding the temperature dependence of ΔG^0_{rxn}.

If $\Delta H^0_{rxn} < 0$ and $\Delta S^0_{rxn} > 0$, then $\underline{\Delta G^0_{rxn} < 0 \text{ at all temperatures}}$.

If $\Delta H^0_{rxn} > 0$ and $\Delta S^0_{rxn} < 0$, then $\underline{\Delta G^0_{rxn} > 0 \text{ at all temperatures}}$.

If $\Delta H^0_{rxn} < 0$ and $\Delta S^0_{rxn} < 0$, then $\underline{\Delta G^0_{rxn} < 0 \text{ at low temperatures}}$.

If $\Delta H^0_{rxn} > 0$ and $\Delta S^0_{rxn} > 0$, then $\underline{\Delta G^0_{rxn} < 0 \text{ at high temperatures}}$.

Practice Test Four
Modules 12-15

Level 1   1. List six strong acids and the eight strong bases.

Level 1   2. Choose all of the following statements that are true:

        a) All Arrhenius bases are also Brønsted-Lowry bases.
        b) All Brønsted-Lowry bases are also Arrhenius bases.
        c) All Arrhenius acids are also Lewis acids.
        d) All Lewis acids are also Arrhenius acids.
        e) All Arrhenius acids are also Brønsted-Lowry acids *and* Lewis acids.

Level 3   3. Determine the *primary* intermolecular force for each molecule.

        a) CH_4
        b) CH_2Cl_2
        c) CH_3COOH
        d) HF
        e) PCl_3

Level 3   4. Arrange the following in order of *decreasing* boiling point:
                CaO, CCl_4, CH_2Br_2, CH_3COOH

Level 1   5. What volume is occupied by a 42.5 g sample of CH_4 gas at 1.34 atm and 32°C?

Level 2   6. Tungsten has a density of 19.3 g/cm^3 and crystallizes in a cubic lattice with a unit cell edge length of 3.16×10^{-10} m. What type of cubic unit cell is formed?

Level 3   7. Determine the %w/w and the $X_{phosphoric\ acid}$ of a 2.75 *m* aqueous solution of H_3PO_4.

Level 1   8. The freezing point of an aqueous solution containing 15 g of a nonelectrolyte in 150 mL of water is -5.4 °C. What is the molecular weight of the compound. K_f for water is 1.86 °C/m.

Level 1   9. Calculate the amount of heat evolved in the conversion of 21.3 grams of steam at 230.0°C to ice at -12.6°C. The specific heats are: $H_2O(s)$ = 2.09 J/g·°C; $H_2O(\ell)$ = 4.18 J/g·°C; $H_2O(g)$ = 2.03 J/g·°C. The heats of fusion and vaporization are: $H_2O(s)$ = 333 J/g; $H_2O(\ell)$ = 2260 J/g, respectively.

Level 1   10. Using the table of thermodynamic data provided, calculate the ΔH^0_{rxn} for the following reaction. Determine whether the reaction is endothermic or exothermic.

$$SiH_4(g) + 2\ O_2(g) \rightarrow SiO_2(s) + 2\ H_2O(\ell)$$

Species	ΔH^0_f (kJ/mol)
SiH_4	34.3
SiO_2	-910.9
H_2O	-285.8

Module 16 Predictor Questions

The following questions may help you to determine to what extent you need to study this module. The questions are ranked according to ability.

Level 1 = basic proficiency
Level 2 = mid level proficiency
Level 3 = high proficiency

If you can correctly answer the Level 3 questions, then you probably do not need to spend much time with this module. If you are only able to answer the Level 1 problems, then you should review the topics covered in this module.

Level 2   1. How are the rates of the disappearance of O_3 and the appearance of C_2H_4O and O_2 related to the disappearance of C_2H_4 in the following reaction?

$$C_2H_4(g) + O_3(g) \rightarrow C_2H_4O\,(g) + O_2(g)$$

Level 1   2. Determine the rate-law expression for this reaction using the experimental data provided.

$$2\,A + B_2 + C \rightarrow A_2B + BC$$

Trial	Initial [A]	Initial [B$_2$]	Initial [C]	Initial rate of formation of BC
1	0.20 M	0.20 M	0.20 M	2.4 x 10^{-6} $M{\cdot}min^{-1}$
2	0.40 M	0.30 M	0.20 M	9.6 x 10^{-6} $M{\cdot}min^{-1}$
3	0.20 M	0.30 M	0.20 M	2.4 x 10^{-6} $M{\cdot}min^{-1}$
4	0.20 M	0.40 M	0.40 M	4.8 x 10^{-6} $M{\cdot}min^{-1}$

Level 1   3. The second order reaction $2\,CH_4 \rightarrow C_2H_2 + 3\,H_2$ has a rate constant of 5.76 $M^{-1}{\cdot}min^{-1}$ at 1600 K. How long, in min, would it take for the concentration of CH_4 to be reduced from 0.89 M to 5.25 x 10^{-4} M?

Level 1   4. Consider the following rate law expression: rate $= k[A]^2[B]$. Which of the following is not true about the reaction having this expression?

a) The reaction is first order in B.
b) The reaction is overall third order.
c) The reaction is second order in A.
d) A and B must both be reactants.
e) Doubling the concentration of A doubles the rate.

Level 1   5. The following reaction is first order with respect to CS_2 and has $k = 2.8 \times 10^{-7}$ s^{-1} at 1000°C.
$$CS_2 \rightarrow CS + S$$

If the initial concentration of CS_2 is 2.0 M, what will the concentration of CS_2 be 28 days after the reaction begins?

Level 1   6. Consider again the decomposition reaction of CS_2 (using the same k value given above). How many days will pass before a 10.0 g sample of CS_2 decomposes to the point that only 2.00 g of CS_2 remains?

Level 1   7. What is the half-life, in days, of the CS_2 decomposition reaction described in question 6?

Level 1   8. The following reaction is second order with $k = 0.0442$ M^{-1}s^{-1}.
$$2\ C_2F_4 \rightarrow C_4F_8$$

If the initial concentration of C_2F_4 is 0.0675 M, what will be the concentration of C_2F_4 100. seconds after the reaction begins?

What is the half-life of the reaction?

Level 2   9. Find E_a for a reaction in which the rate constant quadruples as the temperature is increased from 298 K to 318 K.

Module 16 Predictor Question Solutions

1. $\dfrac{-\Delta[C_2H_4]}{\Delta t} = \dfrac{-\Delta[O_3]}{\Delta t} = \dfrac{+\Delta[C_2H_4O]}{\Delta t} = \dfrac{+\Delta[O_2]}{\Delta t}$

2. When [A] is doubled (from trial 3 to trial 2) but [B] and [C] are held constant, the rate increases 4x. The reaction is thus second order with respect to A.

   When [B] is increased by half (from trial 1 to trial 3) but [A] and [C] are held constant, the rate does not change. The reaction is thus zero order with respect to B.

   When [C] is doubled (from trial 3 to trial 4) but [A] is held constant ([B] changes, but it has already been determined that this does not affect the rate), the rate doubles. The reaction is thus first order with respect to C.

   The rate law expression is: rate $= k[A]^2[B]^0[C]^1$

3.

   $\dfrac{1}{[A]} - \dfrac{1}{[A_0]} = kt$

   $\dfrac{1}{5.24 \times 10^{-4}\ M} - \dfrac{1}{0.89\ M} = (5.76\ M^{-1}\text{min}^{-1})t$

   $t = 330\ \text{min}$

4. Statement e) is not true. The reaction is second order with respect to A, so doubling [A] increases the rate 4x rather than doubling it.

5.

   $(28\ \text{days})\left(\dfrac{24\ \text{h}}{1\ \text{day}}\right)\left(\dfrac{60\ \text{min}}{1\ \text{h}}\right)\left(\dfrac{60\ \text{sec}}{1\ \text{min}}\right) = 2.42 \times 10^6\ \text{s}$

   $[A] = [A_0]e^{-kt}$

   $[A] = (2.0\ M)e^{-(2.8 \times 10^{-7}s^{-1})(2.42 \times 10^6 s)} = 1.02\ M$

6.

   $[A] = [A_0]e^{-kt} \Rightarrow \dfrac{\ln\dfrac{[A]}{[A_0]}}{-k} = t$

   $t = \dfrac{\ln\dfrac{(2.00\ g)}{(10.0\ g)}}{-2.8 \times 10^{-7} s^{-1}} = 5.75 \times 10^6\ \text{s} = 66.6\ \text{days}$

7.

$$kt_{1/2} = 0.693 \Rightarrow t_{1/2} = \frac{0.693}{k}$$

$$t1/2 = \frac{0.693}{2.8 \times 10^{-7}\ s^{-1}} = 2.48 \times 10^6\ s$$

8.

$$\frac{1}{[A]} - \frac{1}{[A_0]} = kt$$

$$\frac{1}{[A]} = \frac{1}{[A_0]} + kt = \frac{1}{(0.0675\ M)} + (0.0442\ M^{-1}s^{-1})(100.s) = 19.23\ M^{-1}$$

$$[A] = 0.0520\ M$$

$$kt_{1/2} = \frac{1}{[A_0]}$$

$$(0.0442\ M^{-1}s^{-1})t_{1/2} = \frac{1}{(0.0675 M)}$$

$$t_{1/2} = 335\ s$$

9.

$$\ln \frac{k_2}{k_1} = \frac{E_a}{R}\left(\frac{1}{T_1} - \frac{1}{T_2}\right)$$

$$\frac{\ln(4)}{\left(\frac{1}{298\ K} - \frac{1}{318\ K}\right)} = \frac{E_a}{8.314\ \dfrac{J}{mol \cdot K}}$$

$$E_a = 5.46 \times 10^4\ J/mol$$

Module 16
Chemical Kinetics

Introduction

Chemical kinetics describes how quickly chemical reactions occur. There are several factors that chemists can control in order to change the rate of a reaction. These include temperature and the concentrations of the reactants. This module describes:

1. the relationship of the rates of one reactant to the rates of other reactants and products
2. how to determine the order of a reactant from experimental data
3. integrated rate laws for first and second order reactions
4. the effect of temperature on the rate of a reaction using the Arrhenius equation

---

Module 16 Key Equations & Concepts

1. $\text{rate} \propto \dfrac{-\Delta[A]}{\Delta t} = \dfrac{-\Delta[B]}{b\Delta t} = \dfrac{+\Delta[C]}{c\Delta t}$ **for the reaction** $A + bB \rightarrow cC$

    This is the definition of the rate of a reaction based on the concentrations of the reactants or products. (The symbol [A] represents the molar concentration of substance A and similarly for [B] and [C].) Notice that the concentrations of A and B (reactants) decrease with time (-), t, and that of C (a product) increases with time (+).

2. $[A] = [A_0]e^{-kt}$

    This is the integrated rate law for chemical reactions that obey *first order kinetics*. It is used to determine *either* the concentration of a reactant a certain amount of time after a reaction has started *or* the amount of time required for the concentration of a reactant to reach a specified amount. [A] is the concentration of A after time has passed, $[A_0]$ is initial concentration of A, k is the rate constant, and t is the amount of time.

3. $k\, t_{1/2} = 0.693$

    The half-life relationship for *first order reactions* is a method to determine the half-life of a first order reaction given the rate constant or vice versa. The half-life is the length of time for the concentration to reach one-half the initial amount.

4. $\dfrac{1}{[A]} - \dfrac{1}{[A_0]} = kt$

    This is the integrated rate law for chemical reactions that obey *second order kinetics*. It is used to determine *either* the concentration of a reactant a certain amount of time after a reaction has started *or* the amount of time required for the concentration of a reactant to reach a specified amount.

5. $k\, t_{1/2} = \dfrac{1}{[A_0]}$

    The half-life relationship for *second order reactions* is a method to determine the half-life of a second order reaction given the rate constant or vice versa.

147

6. $$\ln\frac{k_2}{k_1} = \frac{E_a}{R}\left(\frac{1}{T_1} - \frac{1}{T_2}\right)$$

The **Arrhenius equation** describes how the rate of a reaction changes when the reaction temperature is increased or decreased. E_a is the activation energy of the reaction; k_2 and k_1 are the rate constants at temperatures T_1 and T_2; R is the universal gas constant.

Sample Exercises
Rates of a Reaction Based on the Concentrations of Products and Reactants

1. *How are the rates of the disappearance of O_2 and the appearance of H_2O related to the rate of disappearance of H_2 in this reaction?*

$$2\,H_2(g) \;+\; O_2(g) \;\rightarrow\; 2\,H_2O(g)$$

The correct answer is $\dfrac{-\Delta[H_2]}{\Delta t} = \dfrac{-2\Delta[O_2]}{\Delta t} = \dfrac{+\Delta[H_2O]}{\Delta t}$.

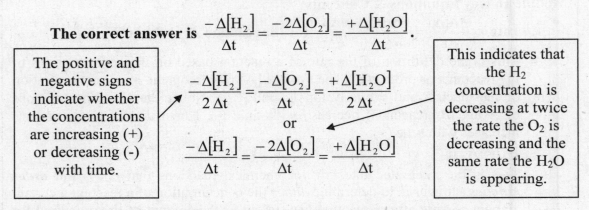

The positive and negative signs indicate whether the concentrations are increasing (+) or decreasing (-) with time.

$$\frac{-\Delta[H_2]}{2\,\Delta t} = \frac{-\Delta[O_2]}{\Delta t} = \frac{+\Delta[H_2O]}{2\,\Delta t}$$

or

$$\frac{-\Delta[H_2]}{\Delta t} = \frac{-2\Delta[O_2]}{\Delta t} = \frac{+\Delta[H_2O]}{\Delta t}$$

This indicates that the H_2 concentration is decreasing at twice the rate the O_2 is decreasing and the same rate the H_2O is appearing.

Determination of the Order of a Reaction from Experimental Data

2. *The following experimental data were obtained for the chemical reaction:*

$$(C_2H_5)_3N + C_2H_5Br \rightarrow (C_2H_5)_4NBr$$

Experiment	$[(C_2H_5)_3N]$ (*M*)	$[C_2H_5Br]$ (*M*)	Relative rate (*M*/min)
1	0.10	0.10	3.0
2	0.20	0.10	6.0
3	0.10	0.30	9.0

What is the rate equation for this reaction and the value of the rate constant, k?
The correct answer is: rate = k $[(C_2H_5)_3N]^1$ $[C_2H_5Br]^1$ and k = 3.0 x 10^2 M^{-1} min^{-1}.

Problems of this type present a set of data in which the concentration of one of the reactants changes while the other reactants' concentrations remain constant. For example, compare experiments 1 and 2. Notice that the $[(C_2H_5)_3N]$ doubles, 0.10 *M* to 0.20 *M*, and the $[C_2H_5Br]$ remains constant at 0.10 *M*. Thus the concentration effects on the rate have been isolated to the $[(C_2H_5)_3N]$. Now, look at the relative rates for experiments 1 and 2 which changes from 3.0 to 6.0 *M*/min, i.e. it also doubles. This indicates that the reaction is 1st order with respect to $[(C_2H_5)_3N]$.

Now compare experiments 1 and 3 where the $[(C_2H_5)_3N]$ remains constant at 0.10 M and the $[C_2H_5Br]$ triples, from 0.10 M to 0.30 M. Notice that the rate also triples from 3.0 to 9.0 M/min. Thus we can conclude that this reaction is 1st order with respect to $[C_2H_5Br]$.

That is the information required to write the form of the rate law for this reaction. Thus we can conclude that this reaction is 1st order with respect to $[C_2H_5Br]$. That is the information required to write the form of the rate law for this reaction.

$$\text{rate} = \text{k } [(C_2H_5)_3N]^1 \, [C_2H_5Br]^1$$

We say that this reaction is 1st order with respect to $[(C_2H_5)_3N]$, 1st order with respect to $[C_2H_5Br]$ and 2nd order overall.
(The overall order is the sum of the individual orders.)

CAUTION

A very common mistake is to assume that the order of a reaction is determined by the stoichiometric coefficients of the balanced chemical reaction. This is not correct. **The only method to determine the order of a reaction is analysis of experimental data just as is done in this problem.**

The value of the rate constant, k, can be determined from the data from experiments 1, 2, or 3.

The values of the rate and the concentrations of both $(C_2H_5)_3N$ and C_2H_5Br come from experiment 3.

As an example let us choose experiment 3's data.

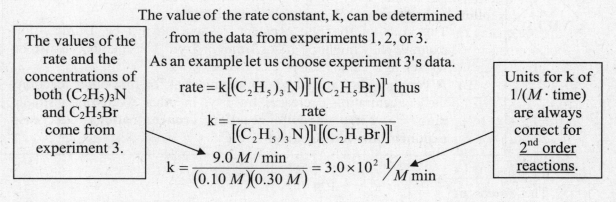

$$\text{rate} = k\left[(C_2H_5)_3 N\right]^1 \left[(C_2H_5Br)\right]^1 \text{ thus}$$

$$k = \frac{\text{rate}}{\left[(C_2H_5)_3 N\right]^1 \left[(C_2H_5Br)\right]^1}$$

$$k = \frac{9.0 \, M/\min}{(0.10 \, M)(0.30 \, M)} = 3.0 \times 10^2 \, \frac{1}{M \, \min}$$

Units for k of $1/(M \cdot \text{time})$ are always correct for 2nd order reactions.

First Order Integrated Rate Law

3. *The following reaction is first order with respect to $[NH_2NO_2]$ and the value of the rate constant, k, is $9.3 \times 10^{-5} \, s^{-1}$. If the initial $[NH_2NO_2] = 2.0$ M, what will the $[NH_2NO_2]$ be 30.0 minutes after the reaction has started?*

$$NH_2NO_2(aq) \; \rightarrow \; N_2O(g) \; + \; H_2O(\ell)$$

The correct answer is: 1.7 M.

$$30.0 \, \text{min} \left(\frac{60.0 \, \text{s}}{\text{min}} \right) = 1800 \, \text{s}$$

We use this equation because the reaction is 1st order with respect to [NH₂NO₂].

Time must be converted from minutes to s because k is in units of s⁻¹.

$$[A] = [A_0] e^{-kt}$$

Rate constant, k and time, t.

Initial concentration of NH₂NO₂(aq).

$$[A] = 2.0 \, M \, e^{-\left(9.3 \times 10^{-5} \, s^{-1}\right) 1800 \, s}$$

$$[A] = 2.0 \, M \, e^{-0.17}$$

$$[A] = 2.0 \, M \, (0.85) = 1.7 \, M$$

After 30.0 minutes the concentration has dropped from 2.0 M to 1.7 M.

4. The following reaction is first order with respect to [NH₂NO₂] and the value of the rate constant, k, is 9.3 x 10⁻⁵ s⁻¹. If the initial [NH₂NO₂] = 2.0 M, how long will it be before the [NH₂NO₂] = 1.5 M?

$$NH_2NO_2(aq) \rightarrow N_2O(g) + H_2O(\ell)$$

The correct answer is: 3.1 x 10³ s or 52 min.

INSIGHT: This problem is a slight variation of exercise 3. All that is required is a little algebra to solve the integrated rate law for t instead of A.

YIELD For reactions that obey simple first order kinetics, i.e. rate = k [A]¹, the following important points must be remembered:

1) The units of the rate constant, k, will always be 1/time. For example they might be 1/s or 1/min or 1/yr. These units can also be written as s⁻¹, min⁻¹, or yr⁻¹.

2) A very common mistake is to assume that for first order reactions the concentration decreases linearly, in other words as a simple ratio. **First order reaction concentrations decrease exponentially not linearly!**

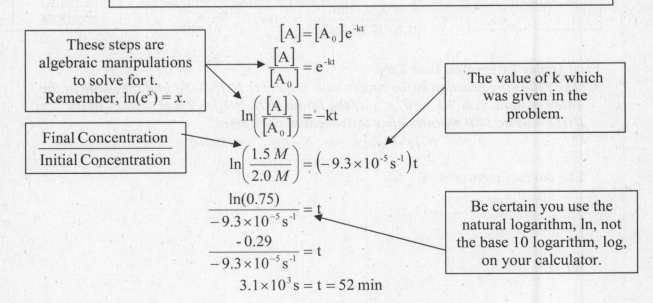

These steps are algebraic manipulations to solve for t. Remember, ln(eˣ) = x.

$$[A] = [A_0] e^{-kt}$$

$$\frac{[A]}{[A_0]} = e^{-kt}$$

The value of k which was given in the problem.

$$\ln\left(\frac{[A]}{[A_0]}\right) = -kt$$

Final Concentration / Initial Concentration

$$\ln\left(\frac{1.5 \, M}{2.0 \, M}\right) = \left(-9.3 \times 10^{-5} \, s^{-1}\right) t$$

$$\frac{\ln(0.75)}{-9.3 \times 10^{-5} \, s^{-1}} = t$$

Be certain you use the natural logarithm, ln, not the base 10 logarithm, log, on your calculator.

$$\frac{-0.29}{-9.3 \times 10^{-5} \, s^{-1}} = t$$

$$3.1 \times 10^3 \, s = t = 52 \, min$$

5. *The following reaction is first order with respect to [NH₂NO₂] and the value of the rate constant, k, is 9.3 x 10⁻⁵ s⁻¹. What is the half-life of this reaction?*

$$NH_2NO_2(aq) \rightarrow N_2O(g) + H_2O(\ell)$$

The correct answer is: 7.5 x 10³ s or 1.2 x 10² min.

$$kt_{1/2} = 0.693$$

$$t_{1/2} = \frac{0.693}{k}$$

$$t_{1/2} = \frac{0.693}{9.3 \times 10^{-5}\,s^{-1}}$$

$$t_{1/2} = 7.5 \times 10^3\,s = 1.2 \times 10^2\,min$$

Use this equation because the reaction is 1ˢᵗ order with respect to [NH₂NO₂].

Second Order Integrated Rate Law

6. *The following reaction at 400.0 K is second order with respect to [CF₃] and the value of the rate constant, k, is 2.51 x 10¹⁰ M⁻¹s⁻¹. If the initial [CF₃] = 2.0 M, what will the [CF₃] be 4.25 x 10⁻¹⁰ seconds after the reaction has started?*

$$2\,CF_3(g) \rightarrow C_2F_6(g)$$

The correct answer is: 8.96 x 10⁻² M.

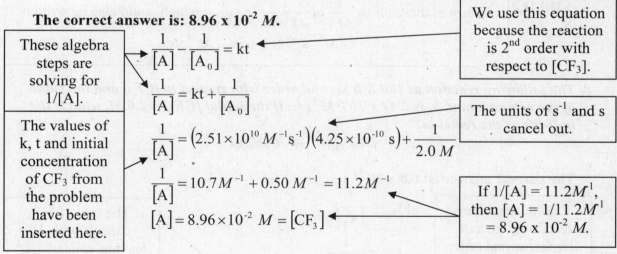

These algebra steps are solving for 1/[A].
The values of k, t and initial concentration of CF₃ from the problem have been inserted here.

$$\frac{1}{[A]} - \frac{1}{[A_0]} = kt$$

$$\frac{1}{[A]} = kt + \frac{1}{[A_0]}$$

$$\frac{1}{[A]} = \left(2.51 \times 10^{10}\,M^{-1}s^{-1}\right)\left(4.25 \times 10^{-10}\,s\right) + \frac{1}{2.0\,M}$$

$$\frac{1}{[A]} = 10.7\,M^{-1} + 0.50\,M^{-1} = 11.2\,M^{-1}$$

$$[A] = 8.96 \times 10^{-2}\,M = [CF_3]$$

We use this equation because the reaction is 2ⁿᵈ order with respect to [CF₃].

The units of s⁻¹ and s cancel out.

If 1/[A] = 11.2M⁻¹, then [A] = 1/11.2M⁻¹ = 8.96 x 10⁻² M.

7. *The following reaction at 400.0 K is second order with respect to [CF₃] and the value of the rate constant, k, is 2.51 x 10¹⁰ M⁻¹s⁻¹. If the initial [CF₃] = 2.0 M, how long will it be before the [CF₃] = 1.5 M?*

$$2\,CF_3(g) \rightarrow C_2F_6(g)$$

The correct answer is: 6.8 x 10⁻¹² s.

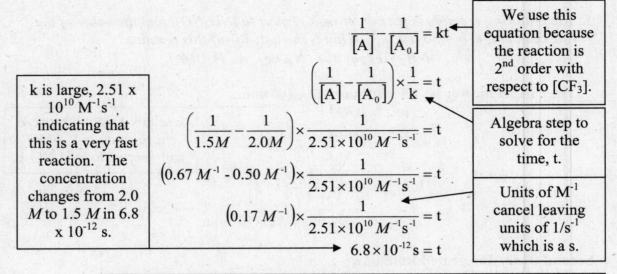

$$\frac{1}{[A]} - \frac{1}{[A_0]} = kt$$

We use this equation because the reaction is 2nd order with respect to $[CF_3]$.

$$\left(\frac{1}{[A]} - \frac{1}{[A_0]}\right) \times \frac{1}{k} = t$$

Algebra step to solve for the time, t.

k is large, $2.51 \times 10^{10}\ M^{-1}s^{-1}$, indicating that this is a very fast reaction. The concentration changes from 2.0 M to 1.5 M in 6.8×10^{-12} s.

$$\left(\frac{1}{1.5M} - \frac{1}{2.0M}\right) \times \frac{1}{2.51 \times 10^{10}\ M^{-1}s^{-1}} = t$$

$$\left(0.67\ M^{-1} - 0.50\ M^{-1}\right) \times \frac{1}{2.51 \times 10^{10}\ M^{-1}s^{-1}} = t$$

$$\left(0.17\ M^{-1}\right) \times \frac{1}{2.51 \times 10^{10}\ M^{-1}s^{-1}} = t$$

$$6.8 \times 10^{-12}\ s = t$$

Units of M^{-1} cancel leaving units of $1/s^{-1}$ which is a s.

YIELD

For reactions that obey simple second order kinetics, i.e. rate = k $[A]^2$, the units of k will always be $\dfrac{1}{(\text{concentration})(\text{time})}$. For example, k could be in any of these units: $\dfrac{1}{M\,s}$ or $\dfrac{1}{M\,\min}$ or $\dfrac{1}{M\,yr}$ which could also be written as $M^{-1}\,s^{-1}$, $M^{-1}\,\min^{-1}$, $M^{-1}\,yr^{-1}$.

8. **The following reaction at 400 K is second order with respect to [CF₃] and the value of the rate constant, k, is $2.51 \times 10^{10}\ M^{-1}s^{-1}$. If the initial [CF₃] = 2.0 M, what is the half-life of the reaction?**

$$2\ CF_3(g) \rightarrow C_2F_6(g)$$

The correct answer is: 6.8×10^{-12} s.

Unlike first order reactions, the half-life for second order reactions changes with the initial concentration of the reactant.

$$kt_{1/2} = \frac{1}{[A_0]}$$

$$t_{1/2} = \frac{1}{k[A_0]}$$

$$t_{1/2} = \frac{1}{\left(2.51 \times 10^{10}\ M^{-1}s^{-1}\right)(2.0M)}$$

$$t_{1/2} = \frac{1}{5.0 \times 10^{10}\ s^{-1}} = 2.0 \times 10^{-11}\ s$$

The unit of M^{-1} cancels with M leaving units of $1/s^{-1}$ = 1 s.

Arrhenius Equation

9. *A reaction has an activation energy of 52.0 kJ/mol and a rate constant, k, of 7.50 x 10^2 s^{-1} at 300.0 K. What is the rate constant for this reaction at 350.0 K?*

The correct answer is: **1.42 x 10^4 s^{-1}.**

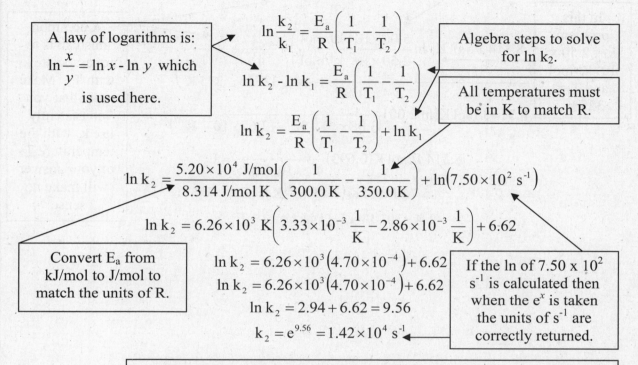

A law of logarithms is:
$$\ln \frac{x}{y} = \ln x - \ln y \text{ which}$$
is used here.

$$\ln \frac{k_2}{k_1} = \frac{E_a}{R}\left(\frac{1}{T_1} - \frac{1}{T_2}\right)$$

Algebra steps to solve for $\ln k_2$.

$$\ln k_2 - \ln k_1 = \frac{E_a}{R}\left(\frac{1}{T_1} - \frac{1}{T_2}\right)$$

All temperatures must be in K to match R.

$$\ln k_2 = \frac{E_a}{R}\left(\frac{1}{T_1} - \frac{1}{T_2}\right) + \ln k_1$$

$$\ln k_2 = \frac{5.20 \times 10^4 \text{ J/mol}}{8.314 \text{ J/mol K}}\left(\frac{1}{300.0 \text{ K}} - \frac{1}{350.0 \text{ K}}\right) + \ln\left(7.50 \times 10^2 \text{ s}^{-1}\right)$$

Convert E_a from kJ/mol to J/mol to match the units of R.

$$\ln k_2 = 6.26 \times 10^3 \text{ K}\left(3.33 \times 10^{-3} \frac{1}{K} - 2.86 \times 10^{-3} \frac{1}{K}\right) + 6.62$$

$$\ln k_2 = 6.26 \times 10^3\left(4.70 \times 10^{-4}\right) + 6.62$$

$$\ln k_2 = 6.26 \times 10^3\left(4.70 \times 10^{-4}\right) + 6.62$$

$$\ln k_2 = 2.94 + 6.62 = 9.56$$

$$k_2 = e^{9.56} = 1.42 \times 10^4 \text{ s}^{-1}$$

If the ln of 7.50 x 10^2 s^{-1} is calculated then when the e^x is taken the units of s^{-1} are correctly returned.

INSIGHT: Kinetics problems that deal with changing rates, or rate constants (k), and temperature changes require use of the Arrhenius equation.

10. *What is the activation energy of a reaction that has a rate constant of 2.50 x 10^2 kJ/mol at 325K and a rate constant of 5.00 x 10^2 kJ/mol at 375 K?*
The correct answer is: **14 kJ/mol.**

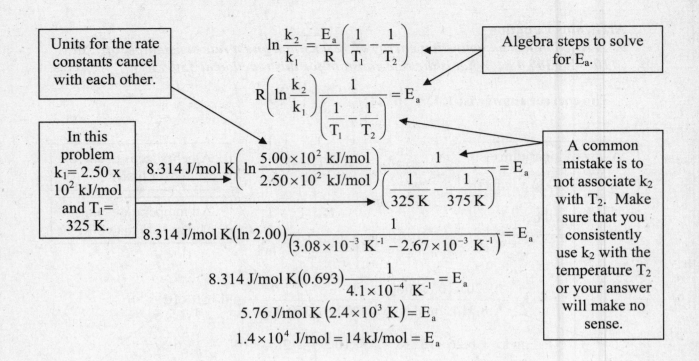

Units for the rate constants cancel with each other.

Algebra steps to solve for E_a.

$$\ln \frac{k_2}{k_1} = \frac{E_a}{R}\left(\frac{1}{T_1} - \frac{1}{T_2}\right)$$

$$R\left(\ln \frac{k_2}{k_1}\right)\frac{1}{\left(\dfrac{1}{T_1} - \dfrac{1}{T_2}\right)} = E_a$$

In this problem k_1= 2.50 x 10^2 kJ/mol and T_1= 325 K.

A common mistake is to not associate k_2 with T_2. Make sure that you consistently use k_2 with the temperature T_2 or your answer will make no sense.

$$8.314 \text{ J/mol K}\left(\ln \frac{5.00 \times 10^2 \text{ kJ/mol}}{2.50 \times 10^2 \text{ kJ/mol}}\right)\frac{1}{\left(\dfrac{1}{325 \text{ K}} - \dfrac{1}{375 \text{ K}}\right)} = E_a$$

$$8.314 \text{ J/mol K}(\ln 2.00)\frac{1}{\left(3.08 \times 10^{-3} \text{ K}^{-1} - 2.67 \times 10^{-3} \text{ K}^{-1}\right)} = E_a$$

$$8.314 \text{ J/mol K}(0.693)\frac{1}{4.1 \times 10^{-4} \text{ K}^{-1}} = E_a$$

$$5.76 \text{ J/mol K}\left(2.4 \times 10^3 \text{ K}\right) = E_a$$

$$1.4 \times 10^4 \text{ J/mol} = 14 \text{ kJ/mol} = E_a$$

Module 17 Predictor Questions

The following questions may help you to determine to what extent you need to study this module. The questions are ranked according to ability.

Level 1 = basic proficiency
Level 2 = mid level proficiency
Level 3 = high proficiency

If you can correctly answer the Level 3 questions, then you probably do not need to spend much time with this module. If you are only able to answer the Level 1 problems, then you should review the topics covered in this module.

Level 1   1. The equilibrium concentrations of the species in the reaction given below are as follows: $[N_2] = 0.301$ M, $[H_2] = 0.240$ M, and $[NH_3] = 0.0541$ M. What is the value of K_c for this reaction?

$$N_2(g) + 3 H_2(g) \leftrightarrows 2 NH_3(g)$$

Level 2   2. At a certain temperature, $K_c = 14.5$ for the reaction below:
$$CO(g) + 2 H_2(g) \leftrightarrows CH_3OH(g)$$

If the equilibrium concentrations of CO and CH_3OH are 1.029 M and 1.86 M, respectively, then what is the equilibrium concentration of H_2?

Level 1   3. Some nitrogen and hydrogen gases are pumped into an empty 5.00 L vessel at 500°C. When equilibrium was established, 3.00 moles of N_2, 2.10 moles of H_2, and 0.298 moles of NH_3 were present. What is the value of K_c at 500 °C for this reaction? See the reaction above.

Level 3   4. Given that: $PCl_5(g) \rightleftarrows PCl_3(g) + Cl_2(g)$ has $K_c = 0.040$ at 450°C, what is the equilibrium concentration of $PCl_5(g)$ if 0.20 mol of $PCl_5(g)$ are placed in a 1.00 L container at 450°C? What is the new equilibrium concentration of $PCl_5(g)$ if the container's volume is halved at 450°C?

Level 1   5. Select all of the following stresses that would shift this reaction's equilibrium to the right (favoring the forward reaction).
$$2 NO(g) + Cl_2(g) \rightleftarrows 2 NOCl(g) \ \Delta H < 0$$
a) Add more NOCl.
b) Remove some Cl_2.
c) Lower the temperature.
d) Add more NO.

Level 2   6. Consider the reaction below. Determine which sets of reaction conditions will produce the maximum yield of product.

$$A(g) + B(g) \rightleftarrows D(g) + Heat$$

a) 100°C, 50 atm
b) 100 °C, 10 atm
c) 500 °C, 50 atm, catalyst
d) 100°C, 50 atm, catalyst
e) 500°C, 10 atm, catalyst

Level 1   7. The equilibrium constant, K_c, for the following reaction is 0.0154 at high temperature. A mixture in a container at this temperature has the following concentrations: $[H_2] = 1.11\ M$, $[I_2] = 1.30\ M$, $[HI] = 0.181\ M$. Which of the following statements concerning the reaction and the reaction quotient, Q, is true?

$$H_2(g) + I_2(g) \rightleftarrows 2\,HI(g)$$

a) $Q = K_c$
b) $Q > K_c$; more HI will be produced
c) $Q > K_c$; more H_2 and I_2 will be produced
d) $Q < K_c$; more HI will be produced
e) $Q < K_c$; more H_2 and I_2 will be produced

Level 3   8. Calculate the thermodynamic equilibrium constant at 25°C for a reaction which has a $\Delta G^0 = 11.3$ kJ per mol of reaction.
R = 8.314 J/mol·K

Module 17 Predictor Question Solutions

1. $K_c = \dfrac{[NH_3]^2}{[H_2]^3[N_2]} = \dfrac{[0.0541]^2}{[0.240]^3[0.301]} = 0.703$

2.

$K_c = \dfrac{[CH_3OH]}{[CO][H_2]^2} = 14.5 = \dfrac{[1.86\ M]}{[1.029\ M][H_2]^2}$

$[H_2] = 0.353\ M$

3. $N_2 + 3H_2 \rightleftarrows 2NH_3$

$[NH_3] = \dfrac{0.298\ mol}{5.00\ L} = 0.0596\ M$

$[N_2] = \dfrac{3.00\ mol}{5.00\ L} = 0.600\ M$

$[H_2] = \dfrac{2.10\ mol}{5.00\ L} = 0.420\ M$

$K_p = \dfrac{[NH_3]^2}{[H_2]^3[N_2]} = \dfrac{[0.0596\ M]^2}{[0.420\ M]^3[0.600\ M]} = 0.0799$

4. $PCl_5 \rightleftarrows PCl_3 + Cl_2$

[Initial]	0.20 M	0	0
$\Delta[\ ]$	-x	+x	+x
[Equilibrium]	0.20 M - x	x	x

$K_c = 0.040 = \dfrac{x^2}{0.20 - x} \Rightarrow x^2 + 0.040x - 0.008$

Solve for x using the quadratic equation : $x = 0.0717$

$[PCl_5] = 0.20\ M - 0.0717 = 0.128\ M$

When the volume is halved, the initial concentration of PCl_5 doubles.

$PCl_5 \rightleftarrows PCl_3 + Cl_2$

[Initial]	0.40 M	0	0
$\Delta[\ ]$	-x	+x	+x
[Equilibrium]	0.40 M - x	x	x

$$K_c = 0.040 = \frac{x^2}{0.40 - x} \Rightarrow x^2 + 0.040x - 0.016$$

Solve for x using the quadratic equation : $x = 0.108$

$[PCl_5] = 0.40\ M - 0.108 = 0.292\ M$

5. $2\,NO + Cl_2 \rightleftarrows 2\,NOCl$

The correct answers are c) and d).

6. The correct answers are a) and d).

Catalysts change rates, but they do not alter the position of equilibrium. The presence of a catalyst thus has no effect on product formation other than attaining equilibrium more quickly.

Heat is a product in this exothermic reaction, so increasing temperature is equivalent to adding a product. This shifts equilibrium toward the reactants. Thus, lower temperature is more favorable to product formation.

The reaction occurs in the gas phase. There are more moles of gas on the reactant side of the reaction, so an increase in pressure shifts the equilibrium to the product side.

7. $Q = \dfrac{[HI]^2}{[I_2][H_2]} = \dfrac{[0.181\ M]^2}{[1.30\ M][1.11\ M]} = 0.0227\ M$

Since $Q > K$, the reaction is reactant favored. The correct answer is c).

8.

$\Delta G^0{}_{rxn} = -RT\ln K$

$T = 25^\circ C + 273.15 = 298.15\ K$

$11.3\ kJ/mol = 11.3 \times 10^3\ J/mol$

$\ln K = -\dfrac{\Delta G^0{}_{rxn}}{RT}$

$\ln K = -\dfrac{(11.3 \times 10^3\ J/mol)}{(8.314\dfrac{J}{mol \cdot K})(298.15\ K)}$

$K = 0.0105$

Module 17
Gas Phase Equilibria

Introduction

This module is a description of the basic calculations required for gas phase equilibria problems. Many of the ideas introduced here will be used again in Module 18 with slight variations. This module describes how to:

1. determine the value of K_c, the equilibrium constant, and use it to predict if a reaction is product or reactant favored
2. calculate the concentrations of species in a reaction
3. see how K_P, the equilibrium constant in terms of the partial pressures of the gases, can be calculated and how it is related to K_c
4. use Le Chatelier's Principle and the reaction quotient, Q, to predict the effects of temperature, pressure, and concentration changes on an equilibrium
5. examine the relationship of ΔG and K; and calculate the value of K at different temperatures

---

Module 17 Key Equations & Concepts

Each of the following equations apply to the reaction aA + bB $\leftrightarrow$ cC + dD where a, b, c, and d are the stoichiometric coefficients for the reaction.

1. **The equilibrium constant, $K_c = \dfrac{[C]^c [D]^d}{[A]^a [B]^b}$**

   This equation is used to determine whether the reaction is product or reactant favored (i.e. yields more reactants or products) and to determine the concentrations of the reactants and products at equilibrium. *Concentrations used in K_c must be equilibrium concentrations.*

2. **The equilibrium constant for gas phase reactions, $K_P = \dfrac{(P_C)^c (P_D)^d}{(P_A)^a (P_B)^b}$**

   The equilibrium constant in terms of the partial pressures of the gases, K_P, serves the same purpose as K_c except that the partial pressures of the gases are used instead of the equilibrium concentrations. K_P is used when it is easiest to measure the pressures of the gases instead of the equilibrium concentrations.

3. **The reaction quotient, $Q = \dfrac{[C]^c [D]^d}{[A]^a [B]^b}$**

   The reaction quotient has *the same form as K_c but the concentrations are all nonequilibrium.* Q is used to determine how the position of equilibrium must shift for a nonequilibrium system to attain equilibrium.

4. **$K_P = K_c (RT)^{\Delta n}$ where**

   $\Delta n = \sum$ **moles of gaseous products -** $\sum$ **moles of gaseous reactants**

   This relationship defines how the partial pressure equilibrium constant, K_P, and the equilibrium constant, K_c, are related.

---

5. The Gibbs Free Energy change, $\Delta G^0_{rxn} = -RT \ln K$

K, the thermodynamic equilibrium constant, is related to the standard Gibbs Free Energy change using this relationship.

6. The van't Hoff equation, $\ln\left(\dfrac{K_{T_2}}{K_{T_1}}\right) = \dfrac{\Delta H^0}{R}\left(\dfrac{1}{T_1} - \dfrac{1}{T_2}\right)$

This equation describes how to determine the values of equilibrium constants at different temperatures.

Sample Exercises
Use of the Equilibrium Constant Expression

1. For the following reaction at 298 K, the equilibrium concentrations are [H_2] = 1.50 M, [I_2] = 2.00 M, and [HI] = 3.46 M. What is the value of the equilibrium constant, K_c, for this reaction at 298 K?

$$H_2(g) + I_2(g) \rightleftarrows 2\,HI(g)$$

The correct answer is: 4.00.

Units are not used in equilibrium constants. We are interested in K_c's size.	For this reaction $K_c = \dfrac{[HI]^2}{[H_2][I_2]}$ thus $K_c = \dfrac{[3.46]^2}{[1.50][2.00]} = \dfrac{12.0}{3.00} = 4.00$	A very common mistake is to forget to properly include the stoichiometric coefficients as exponents.

YIELD

The value of an equilibrium constant indicates if the reaction favors the products, the reactants, or both.
1) K_c >10 to 20, the reaction is **product favored**
2) K_c < 1, the reaction is **reactant favored**
3) $1 < K_c < 10$ to 20, the reaction yields a **mixture of reactants and products**

INSIGHT:

Fundamentally, K_c is a ratio of the product concentrations divided by the reactant concentrations. This is why the larger the value of K_c the more product favored the reaction is. K_c is actually defined using a thermodynamic quantity called activity. The activities of gases are the same as their concentrations. In heterogeneous equilibria (those involving gases, liquids, and solids) the activities of the pure solids and liquids are 1 and can be neglected. Thus the K_c for heterogeneous equilibria will only require the gas's concentrations and not any solids or liquids involved in the equilibrium.

2. Consider the following reaction at 298 K: $H_2(g) + I_2(g) \rightleftarrows 2\,HI(g)$

The equilibrium constant is 4.00. If the reaction vessel initially has the following reactant concentrations [H₂] = 6.00 M and [I₂] = 4.00 M. What are the equilibrium concentrations of all species in this reaction?
The correct answer is: [H₂] = 3.6 *M*, [I₂] = 1.6 *M*, and [HI] = 4.8 *M*.

INSIGHT: Problems that give you the K_c and the starting concentrations of the reactants must be solved in this fashion. Setting up a table is a good approach.

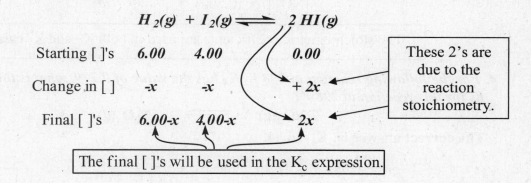

$$H_2(g) + I_2(g) \rightleftharpoons 2\,HI(g)$$

Starting [ ]'s	**6.00**	**4.00**	**0.00**
Change in [ ]	**-x**	**-x**	**+ 2x**
Final [ ]'s	**6.00-x**	**4.00-x**	**2x**

These 2's are due to the reaction stoichiometry.

The final [ ]'s will be used in the K_c expression.

$$K_c = \frac{[HI]^2}{[H_2][I_2]} = 4.00$$

$$= \frac{(2x)^2}{(6.00-x)(4.00-x)} = 4.00$$

Multiplying $(6-x)(4-x)$ gives $24-10x+x^2$.

$$= \frac{4x^2}{24-10x+x^2} = 4.00$$

$$= 4x^2 = 4.00(24-10x+x^2)$$

$$= 4x^2 = 96 - 40x + 4x^2$$

$$= 4x^2 - 4x^2 = 96 - 40x$$

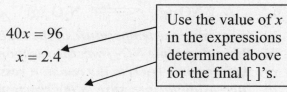

$$40x = 96$$

$$x = 2.4$$

Use the value of x in the expressions determined above for the final [ ]'s.

$$\text{Final}[H_2] = 6.00 - x = 6.00 - 2.4 = 3.6M$$

$$\text{Final}[I_2] = 4.00 - x = 4.00 - 2.4 = 1.6M$$

$$\text{Final}[HI] = 2x = 2(2.4) = 4.8M$$

INSIGHT: In this particular problem the x^2 terms cancel out, simplifying the problem. Problems in which this does NOT occur must be solved using the quadratic equation.

Use of the Equilibrium Constant, K_P

3. For the following reaction at 298 K, the equilibrium partial pressures are:
$P_{NO_2} = 0.500$ atm and $P_{N_2O_4} = 0.0698$ atm. What is the value of K_P for this reaction?

$$2\ NO_2(g) \rightleftharpoons N_2O_4(g)$$

The correct answer is: $K_P = 0.279$.

$$K_P = \frac{[N_2O_4]}{[NO_2]^2} = \frac{0.0698}{(0.500)^2} = \frac{0.0698}{0.250} = 0.279$$

The stoichiometric coefficients are used in both K_C and K_P calculations.

4. For the following reaction at 298 K, K_P has the value of 0.279, what is the value of K_c for this reaction at 298 K?

$$2\ NO_2(g) \rightleftharpoons N_2O_4(g)$$

The correct answer is: $K_c = 6.84$.

$$K_P = K_c (RT)^{\Delta n} \text{ thus } K_c = \frac{K_P}{(RT)^{\Delta n}}$$

$$\Delta n = \sum \text{moles of gaseous products} - \sum \text{moles of gaseous reactants}$$

$$\frac{1}{(0.0821 \times 298)^{-1}} =$$

$$\Delta n = \text{moles of } N_2O_4 - \text{moles of } NO_2 = 1 - 2 = -1$$

$$\frac{1}{(24.5)^{-1}} = 24.5$$

Use the gas law value of R (0.0821 L atm/mol K).

$$K_c = \frac{0.279}{(0.0821 \times 298)^{-1}} = 0.279 \times 24.5 = 6.84$$

Effects of Temperature, Pressure, and Concentration on the Position of Equilibrium

5. What would be the effect of each of these changes on the position of equilibrium of this reaction at 298 K?

$$2\ NO_2(g) \rightleftharpoons N_2O_4(g) \quad \Delta H^0_{rxn} = -57.2\ kJ/mol$$

a) **Increasing the temperature of the reaction**
b) **Removing some NO_2 from the reaction vessel.**
c) **Adding some N_2O_4 to the reaction vessel.**
d) **Increasing the pressure in the reaction vessel by adding an inert gas.**
e) **Decreasing by half the size of the reaction vessel.**
f) **Introducing a catalyst into the reaction vessel.**

The correct answers are the position of equilibrium will shift to the: a) left b) left c) left d) no effect e) right f) no effect.

a) *For exothermic reactions: increasing the temperature shifts the position of equilibrium to the left, decreasing the temperature shifts the position of equilibrium the right. Endothermic reactions behave oppositely.* In this exercise the negative ΔH^0_{rxn} tells us that the reaction is exothermic, thus increasing the temperature shifts the position of equilibrium to the left.

b) *If a reactant's concentration is decreased below the equilibrium concentration, the position of equilibrium will change to restore concentrations that correspond to those predicted by the equilibrium constant.* In this exercise, removing some NO_2 from the reaction vessel decreases the $[NO_2]$. The reaction equilibrium responds to this stress by increasing the $[NO_2]$ and decreasing the $[N_2O_4]$, an equilibrium position shift to the left or reactant side. Adding NO_2 would cause the position of equilibrium to shift to the right or product side.

c) *If a product's concentration is increased above the equilibrium concentration, the equilibrium position will shift to restore concentrations of products and reactants that correspond to those predicted by the equilibrium constant.* Adding some N_2O_4 to the reaction vessel increases the $[N_2O_4]$ above the equilibrium concentration. The reaction equilibrium responds to this stress by decreasing the $[N_2O_4]$ and increasing the $[NO_2]$, an equilibrium position shift to the left or reactant side. Removing N_2O_4 would shift the position of equilibrium to the right or product side.

d) *Adding an inert gas to the reaction mixture has no effect on the equilibrium position because the concentrations of the gases are not changed.* This is a common misconception for students.

e) *If the volume of the reaction vessel is changed, the concentrations of the gases are changed because for gases, $M \propto n/V$. If the vessel's volume is decreased, the equilibrium position will shift to the side that has the fewest moles of gas.* In this exercise the right or product side has the fewest moles.

f) *Adding a catalyst has no effect on the position of equilibrium.* Catalysts change the rates of reactions but not positions of equilibrium.

The Reaction Quotient, Q

6. *A nonequilibrium mixture has a [NO₂] = 0.50 M and [N₂O₄] = 0.50 M at 298 K. How will this reaction respond as equilibrium is reestablished? Will the concentration of the reactants increase or decrease? Will the concentration of the products increase or decrease?*

$$2\ NO_2(g) \rightleftharpoons N_2O_4(g)$$

The correct answer is: the concentration of the products will increase and the concentration of the reactants will decrease until equilibrium is reestablished.

INSIGHT: The reaction quotient, Q, is used to predict how nonequilibrium mixtures will respond as they reestablish equilibrium. Remember, Q uses the nonequilibrium concentrations whereas K_c uses equilibrium concentrations.

$$Q = \frac{[N_2O_4]}{[NO_2]^2} = \frac{[0.50]}{[0.50]^2} = 2.0 \text{ and thus } Q < K_c$$

In exercise 4, we determined that $K_c = 6.84$ for this reaction, thus $Q < K_c$. Because Q is smaller than K_c, this implies that the mixture has too few products and too many reactants. (Q is a fraction and its size is telling us that the numerator needs to be bigger and the denominator smaller to return to equilibrium.) Thus the concentration of the products will increase and the concentration of the reactants will decrease until equilibrium is reestablished.

YIELD

1. If $Q < K_c$, the reaction will consume reactants and yield products to reestablish equilibrium.
2. If $Q > K_c$, the reaction will produce reactants and consume products to reestablish equilibrium.
3. If $Q = K_c$, the reaction is at equilibrium.

Relationship of ΔG^0_{rxn} to the Equilibrium Constant

7. *What is the value of the gaseous equilibrium constant, K_P, at 298 K for this reaction?*

$$H_2(g) + F_2(g) \rightleftharpoons 2\ HF(g)$$

The correct answer is: 5.11 x 10⁹⁵.

We can calculate ΔG^0_{rxn} using the method described in Module 15. For this reaction $\Delta G^0_{rxn} = -546$ kJ/mol or -5.46×10^5 J/mol.

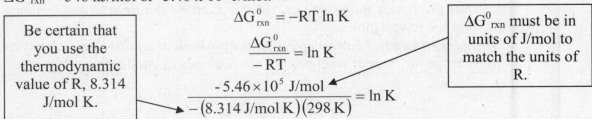

Be certain that you use the thermodynamic value of R, 8.314 J/mol K.

ΔG^0_{rxn} must be in units of J/mol to match the units of R.

$$\Delta G^0_{rxn} = -RT \ln K$$

$$\frac{\Delta G^0_{rxn}}{-RT} = \ln K$$

$$\frac{-5.46 \times 10^5 \text{ J/mol}}{-(8.314 \text{ J/mol K})(298 \text{ K})} = \ln K$$

If ln K = 220, then K $= e^{220}$.	$2.20 \times 10^2 = \ln K$ $e^{2.20 \times 10^2} = K = 5.11 \times 10^{95}$	The size of K indicates that this reaction is definitely product favored.

Evaluation of an Equilibrium Constant at a Different Temperature

8. *The following reaction has an equilibrium constant, K_c, of 6.84 at 298 K. What is the value of K_c at 225K?*

$$2 \, NO_2(g) \quad \rightleftharpoons \quad N_2O_4(g)$$

The correct answer is: 1.15×10^4.

> The problem tells us that $K_{T_1} = 6.84$, $T_1 = 298$ K, and $T_2 = 225$ K. From Module 15 we can calculate that $\Delta H^0 = -57.2$ kJ/mol $= -5.72 \times 10^4$ J/mol.

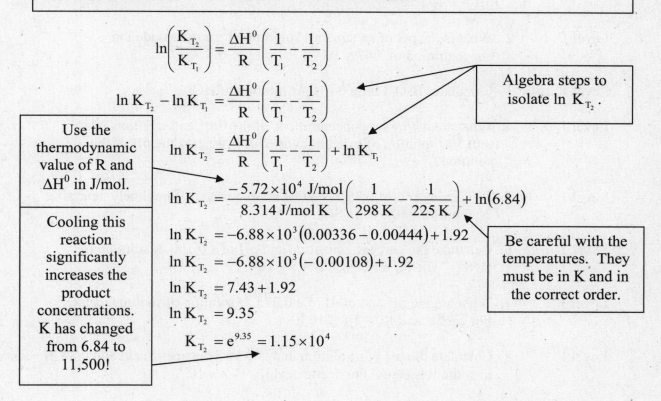

$$\ln\left(\frac{K_{T_2}}{K_{T_1}}\right) = \frac{\Delta H^0}{R}\left(\frac{1}{T_1} - \frac{1}{T_2}\right)$$

$$\ln K_{T_2} - \ln K_{T_1} = \frac{\Delta H^0}{R}\left(\frac{1}{T_1} - \frac{1}{T_2}\right)$$

Algebra steps to isolate $\ln K_{T_2}$.

$$\ln K_{T_2} = \frac{\Delta H^0}{R}\left(\frac{1}{T_1} - \frac{1}{T_2}\right) + \ln K_{T_1}$$

Use the thermodynamic value of R and ΔH^0 in J/mol.

$$\ln K_{T_2} = \frac{-5.72 \times 10^4 \, \text{J/mol}}{8.314 \, \text{J/mol K}}\left(\frac{1}{298 \, \text{K}} - \frac{1}{225 \, \text{K}}\right) + \ln(6.84)$$

$$\ln K_{T_2} = -6.88 \times 10^3 (0.00336 - 0.00444) + 1.92$$

$$\ln K_{T_2} = -6.88 \times 10^3 (-0.00108) + 1.92$$

Be careful with the temperatures. They must be in K and in the correct order.

$$\ln K_{T_2} = 7.43 + 1.92$$

$$\ln K_{T_2} = 9.35$$

Cooling this reaction significantly increases the product concentrations. K has changed from 6.84 to 11,500!

$$K_{T_2} = e^{9.35} = 1.15 \times 10^4$$

Module 18 Predictor Questions

The following questions may help you to determine to what extent you need to study this module. The questions are ranked according to ability.

Level 1 = basic proficiency
Level 2 = mid level proficiency
Level 3 = high proficiency

If you can correctly answer the Level 3 questions, then you probably do not need to spend much time with this module. If you are only able to answer the Level 1 problems, then you should review the topics covered in this module.

Level 1   1. What is the hydroxide ion concentration in an aqueous solution with pH = 6.19?

Level 1   2. What is the pH of an aqueous solution with a hydroxide ion concentration of 5.47×10^{-9}?

Level 1   3. Calculate $[H_3O^+]$ for a 0.010 M solution of HCl.

Level 1   4. What are the molar concentrations of the Ca^{2+} and OH^- ions in a 0.015 M solution of calcium hydroxide? What is the pH of the solution?

Level 1   5. How many mL of 0.35 M NaOH are required to completely neutralize 20.0 mL of 0.026 M H_2SO_4?

Level 2   6. Calculate the percent ionization and pH of a 0.100 M solution of HNO_2. The K_a for HNO_2 is 4.5×10^{-4}.

Level 3   7. What are the pH and pOH of a 0.075 M solution of sodium acetate? For acetic acid $K_a = 1.8 \times 10^{-5}$.

Level 1   8. Calculate the pH of a solution that is 0.10 M in acetic acid and 0.30 M in sodium acetate. For acetic acid $K_a = 1.8 \times 10^{-5}$.

Level 3   9. What ratio of NH_4Cl and NH_3 should be used to give a buffer with pH = 8.50? The K_b for NH_3 is 1.8×10^{-5}.

Level 1   10. What is the correct form of the solubility product constant for $Ba_3(AsO_4)_2$?

Level 1   11. If the solubility of BiI_3 is 7.7×10^{-3} g/L and the solubility of $Fe(OH)_2$ is 1.1×10^{-3} g/L (both in water at 25°C), then what are the values of the K_{sp} for each compound?

Module 18 Predictor Question Solutions

1. $pH = -\log[H^+]$
   $6.19 = -\log[H^+]$
   $[H^+] = 10^{-6.19} = 6.46 \times 10^{-7}\ M$

   $K_w = [OH^-][H^+] = 1.00 \times 10^{-14}\ M$
   $[OH^-][6.46 \times 10^{-7}\ M] = 1.00 \times 10^{-14}\ M$
   $[OH^-] = 1.55 \times 10^{-8}\ M$

2. $pOH = -\log[OH^-] = -\log[5.47 \times 10^{-9}\ M] = 8.26$
   $pOH + pH = 14.00$
   $pH = 14.00 - pOH = 14.00 - 8.26$
   $pH = 5.74$

3. $\left(\dfrac{0.010\ \text{mol HCl}}{1\ \text{L}}\right)\left(\dfrac{1\ \text{mol H}^+}{1\ \text{mol HCl}}\right) = 0.010\ M\ H^+ = 0.010\ M\ H_3O^+$

4.

   $\left(\dfrac{0.015\ \text{mol Ca(OH)}_2}{\text{L}}\right)\left(\dfrac{1\ \text{mol Ca}^{2+}}{1\ \text{mol Ca(OH)}_2}\right) = 0.015\ M\ Ca^{2+}$

   $\left(\dfrac{0.015\ \text{mol Ca(OH)}_2}{\text{L}}\right)\left(\dfrac{2\ \text{mol OH}^-}{1\ \text{mol Ca(OH)}_2}\right) = 0.030\ M\ Ca^{2+}$

   $pOH = -\log[OH^-] = -\log[0.030] = 1.52$
   $pH = 14.00 - pOH = 14.00 - 1.52 = 12.48$

5. The reaction is described by the equation: $H_2SO_4 + 2\ NaOH \rightarrow 2\ H_2O + Na_2SO_4$

   $\left(\dfrac{0.026\ \text{mol H}_2\text{SO}_4}{\text{L}}\right)\left(\dfrac{20.0\ \text{mL}}{}\right)\left(\dfrac{10^{-3}\ \text{L}}{1\ \text{mL}}\right)\left(\dfrac{2\ \text{mol NaOH}}{1\ \text{mol H}_2\text{SO}_4}\right)\left(\dfrac{\text{L}}{0.35\ \text{mol NaOH}}\right)\left(\dfrac{1\ \text{mL}}{10^{-3}\ \text{L}}\right) = 2.97\ \text{mL}$

6. $HNO_2 \rightleftarrows H^+ + NO_2^-$

	HNO_2	H^+	NO_2^-
[Initial]	0.100 M	0	0
$\Delta[\ ]$	-x	+x	+x
[Equilibrium]	0.100 M - x	x	x

167

$$K_a = \frac{[H^+][NO_2^-]}{[HNO_2]} = \frac{x^2}{0.100 - x}$$

Since x is much smaller than 0.100, the equation can be simplfied to : $K_a = \frac{x^2}{0.100} = 4.5 \times 10^{-4}$

$$x = 6.7 \times 10^{-3} = [H^+]$$

$$pH = -\log[H^+] = -\log[6.71 \times 10^{-3}] = 2.17$$

$$\% \text{ ionization} = \frac{[\text{ionized acid}]}{[\text{initial acid}]}(100) = \frac{[6.71 \times 10^{-3}]}{[0.100]}(100) = 6.7\%$$

7. $CH_3COO^- \rightarrow CH_3COOH + OH^-$

[Initial]	0.075 M	0	0
$\Delta[\ ]$	-x	+x	+x
[Equilibrium]	0.075 M - x	X	x

$$K_w = K_a K_b$$

$$K_b = \frac{K_w}{K_a} = \frac{1.00 \times 10^{-14}}{1.8 \times 10^{-5}} = 5.56 \times 10^{-10}$$

$$K_b = 5.56 \times 10^{-10} = \frac{x^2}{0.075 - x} = \frac{x^2}{0.075}$$

$$x = 6.45 \times 10^{-6} = [OH^-]$$

$$pOH = -\log[OH^-] = -\log[6.45 \times 10^{-6}] = 5.19$$

$$pH = 14.00 - pOH = 14.00 - 5.19 = 8.81$$

8. This is a buffer problem. Use the Henderson-Hasselbalch equation. Since the buffer solution contains a weak acid and its conjugate base, use the pK_a.

$$K_a = 1.8 \times 10^{-5}$$

$$pK_a = -\log(K_a) = 4.74$$

$$pH = pKa + \log\frac{[\text{conjugate base}]}{[\text{acid}]} = 4.74 + \log\frac{[0.30\ M]}{[0.10\ M]} = 5.22$$

9. This is a buffer problem. Use the Henderson-Hasselbalch equation. Since the buffer solution contains a weak base and its conjugate acid, use the pK_b.

$K_b = 1.8 \times 10^{-5}$

$pK_b = -\log(K_b) = 4.74$

$pOH = 14.00 - pH = 14.00 - 8.50 = 5.50$

$pOH = 5.50 = pK_b + \log\dfrac{[\text{conjugate base}]}{[\text{acid}]} = 4.74 + \log\dfrac{[\text{conjugate base}]}{[\text{acid}]}$

$5.75 = \dfrac{[\text{conjugate base}]}{[\text{acid}]}$

10. $Ba_3(AsO_4)_2 \rightleftharpoons 3Ba^{2+} + 2AsO_4^{2-}$
    $K_{sp} = [Ba^{2+}]^3[AsO_4^{2-}]^2$

11.

$\left(\dfrac{7.7 \times 10^{-3}\,\text{g BiI}_3}{L}\right)\left(\dfrac{1\,\text{mol BiI}_3}{589.68\,\text{g BiI}_3}\right) = 1.31 \times 10^{-5}\,M$

$BiI_3 \rightleftharpoons Bi^{3+} + 3I^-$

$Bi^{3+} = 1.31 \times 10^{-5}\,M$

$[I^-] = 3(1.31 \times 10^{-5}\,M) = 3.93 \times 10^{-5}\,M$

$K_{sp} = [Bi^{3+}][I^-]^3 = [1.31 \times 10^{-5}\,M][3.93 \times 10^{-5}\,M]^3 = 7.95 \times 10^{-19}$

$\left(\dfrac{1.1 \times 10^{-3}\,\text{g Fe(OH)}_2}{L}\right)\left(\dfrac{1\,\text{mol Fe(OH)}_2}{89.87\,\text{g Fe(OH)}_2}\right) = 1.22 \times 10^{-5}\,M$

$Fe(OH)_2 \rightleftharpoons Fe^{2+} + 2OH^-$

$Fe^{2+} = 1.22 \times 10^{-5}\,M$

$[OH^-] = 2(1.22 \times 10^{-5}\,M) = 2.44 \times 10^{-5}\,M$

$K_{sp} = [Fe^{3+}][OH^-]^2 = [1.22 \times 10^{-5}\,M][2.44 \times 10^{-5}\,M]^2 = 7.26 \times 10^{-15}$

Module 18
Aqueous Equilibria

Introduction

This module describes the calculations required to determine the concentrations of species in various types of aqueous solutions. Exercises in this module will include:

1. determining the hydronium and hydroxide ion concentrations in solutions
2. calculating the pH, pOH, and % ionization of various solutions
3. acid-base titration calculations
4. determining the pH and pOH of solutions in hydrolysis
5. calculating the concentrations and pH of buffer solutions
6. calculating the ion concentrations for insoluble solids

Module 18 Key Equations & Concepts.

1. The ionization constant for water,
$$K_w = [H_3O^+][OH^-] = 1.00 \times 10^{-14}$$
$$14 = pH + pOH$$

This is used to calculate either the hydronium or hydroxide ion concentration in aqueous solutions given the concentration of either one of these ions. The second equation is mathematically equivalent to the first and relates the pH and pOH.

$$pH = -\log[H^+]$$
2. $$pOH = -\log[OH^-]$$
$$pK_a = -\log K_a$$

In chemistry, the symbol pX is defined as the $-\log X$. These equations define pH, pOH, and pK_a. pH is a condensed method to write the H^+ or H_3O^+ concentration in aqueous solutions. pOH is an equivalent method of writing the aqueous OH^- concentration. pK_a is the shorthand method for writing Ka values.

For the weak acid equilibrium $HA \rightleftharpoons H^+ + A^-$ $\quad K_a = \dfrac{[H^+][A^-]}{[HA]}$

3. For the weak base equilibrium $BOH \rightleftharpoons B^+ + OH^-$ $\quad K_b = \dfrac{[B^+][OH^-]}{BOH}$

$$\% \text{ ionization} = \dfrac{[\text{ionized species}]}{[\text{initial species}]} \times 100$$

K_a and K_b are ionization constants for weak acids and weak bases, respectively. They are used to determine the concentrations of all species in aqueous solutions of weak acids and bases. The % ionization is also used in weak acid and base calculations to describe how much of the acid or base is ionized in solution. K_a and K_b values are tabulated in an appendix in your textbook.

---

4. The Henderson-Hasselbalch equations:

For an acidic buffer solution $pH = pK_a + \log \dfrac{[salt]}{[acid]}$

For a basic buffer solution $pOH = pK_b + \log \dfrac{[salt]}{[base]}$

These are used to find the pH of buffer solutions given concentrations of the salt and acid for acidic buffers, or the salt and base for basic buffers.

5. For acid - base conjugate pairs $K_w = K_a \times K_b$

In buffer and hydrolysis calculations, this relationship is used to determine the acid ionization constant for the conjugate acid of a weak base or the base ionization constant for the conjugate base of a weak acid.

Sample Exercises
Water Ionization Constant

1. What is the [OH⁻] in an aqueous solution that has a pH of 5.25?
   The correct answer is: **[OH⁻] = 1.8 x 10⁻⁹ M.**

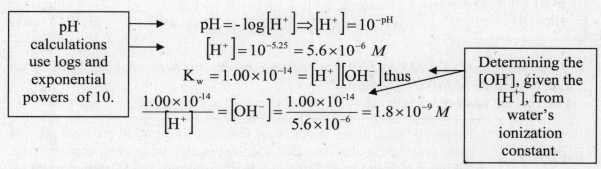

2. What is the pH of an aqueous solution that has a [OH⁻] = 3.45 x 10⁻³?
   The correct answer is: **pH = 11.538.**

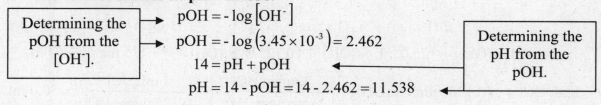

Strong Acid or Base Dissociation

3. What is the pH of an aqueous 0.025 M Sr(OH)₂ solution?
   The correct answer is: **pH = 12.70.**

INSIGHT: Strontium hydroxide, $Sr(OH)_2$, is a water soluble, strong, polyhydroxy base. When it dissociates in water, the $[OH^-]$ in solution will be twice the molarity of the $Sr(OH)_2$. This will also be true for $Ca(OH)_2$ and $Ba(OH)_2$. There is one strong polyprotic acid that you need to be aware of, H_2SO_4. For H_2SO_4, the $[H^+]$ will be twice the molarity.

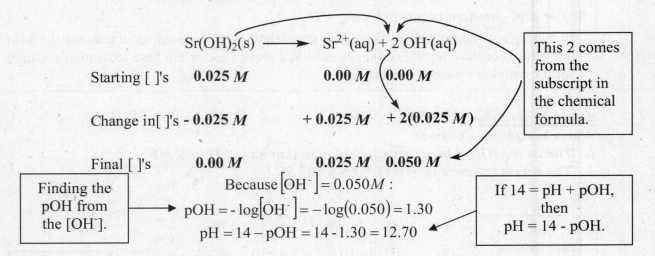

$$Sr(OH)_2(s) \longrightarrow Sr^{2+}(aq) + 2\,OH^-(aq)$$

Starting [ ]'s   **0.025 M**   **0.00 M**   **0.00 M**

Change in[ ]'s **- 0.025 M**   **+ 0.025 M**   **+ 2(0.025 M)**

Final [ ]'s   **0.00 M**   **0.025 M**   **0.050 M**

This 2 comes from the subscript in the chemical formula.

Finding the pOH from the $[OH^-]$.

Because $[OH^-] = 0.050M$:

$$pOH = -\log[OH^-] = -\log(0.050) = 1.30$$
$$pH = 14 - pOH = 14 - 1.30 = 12.70$$

If 14 = pH + pOH, then pH = 14 - pOH.

4. How many mL of 0.125 M HCl are required to exactly neutralize 25.0 mL of an aqueous 0.025 M Sr(OH)₂ solution?
The correct answer is: 50.0 mL.

INSIGHT: The word "neutralize" is your clue that this is a titration problem. In this case, you should also note that it is the reaction of a strong acid with the dihydroxy strong base, $Sr(OH)_2$. In all titrations, the 1st step must be to **write a balanced chemical reaction**.

$$2\,HCl(aq) + Sr(OH)_2\,(aq) \rightarrow SrCl_{2(aq)} + 2\,H_2O(\ell)$$

$$?\ \text{mmol}\ Sr(OH)_2 = \left(25.0\ \text{mL}\ Sr(OH)_2\right)\left(0.125M\ Sr(OH)_2\right) = 3.125\ \text{mmol}\ Sr(OH)_2$$

$$?\,\text{mL}\ HCl = \left(3.125\ \text{mmol}\ Sr(OH)_2\right)\left(\frac{2\ \text{mmol}\ HCl}{1\ \text{mmol}\ Sr(OH)_2}\right)\left(\frac{1\ \text{mL}\ HCl}{0.125\ \text{mmol}\ HCl}\right) = 50.0\ \text{mL}$$

This reaction ratio is important.

M inverted and used as a conversion factor.

Weak Acid or Base Ionization

5. *What is the pH and % ionization of an aqueous 0.125 M acetic acid, CH$_3$COOH, solution?* K$_a$ = 1.8 x 10^{-5}
   The correct answer is: pH = 2.82 and % ionization =1.2%.

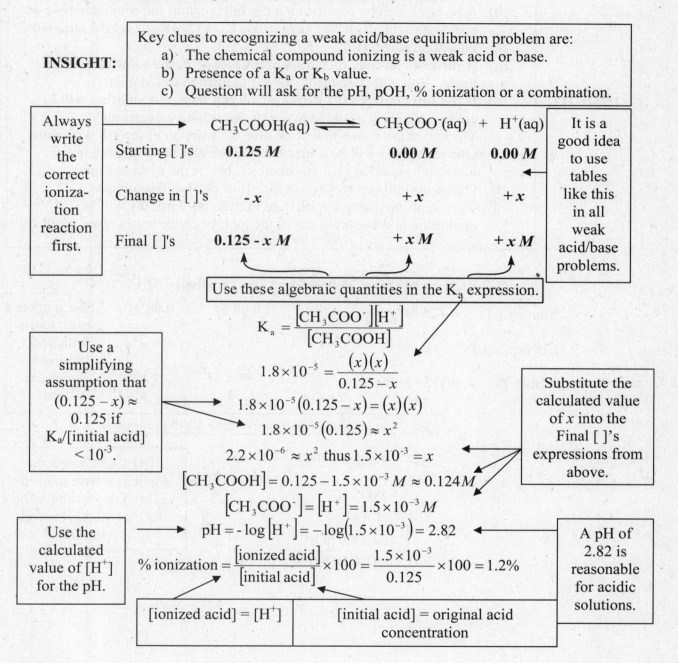

INSIGHT:
Key clues to recognizing a weak acid/base equilibrium problem are:
a) The chemical compound ionizing is a weak acid or base.
b) Presence of a K$_a$ or K$_b$ value.
c) Question will ask for the pH, pOH, % ionization or a combination.

Always write the correct ionization reaction first.

It is a good idea to use tables like this in all weak acid/base problems.

$$CH_3COOH(aq) \rightleftharpoons CH_3COO^-(aq) + H^+(aq)$$

Starting [ ]'s   0.125 M 0.00 M 0.00 M

Change in [ ]'s $-x$ $+x$ $+x$

Final [ ]'s   0.125 - x M $+x$ M $+x$ M

Use these algebraic quantities in the K$_a$ expression.

$$K_a = \frac{[CH_3COO^-][H^+]}{[CH_3COOH]}$$

Use a simplifying assumption that $(0.125 - x) \approx$ 0.125 if K$_a$/[initial acid] < 10^{-3}.

$$1.8 \times 10^{-5} = \frac{(x)(x)}{0.125 - x}$$

$$1.8 \times 10^{-5}(0.125 - x) = (x)(x)$$

$$1.8 \times 10^{-5}(0.125) \approx x^2$$

$$2.2 \times 10^{-6} \approx x^2 \text{ thus } 1.5 \times 10^{-3} = x$$

Substitute the calculated value of x into the Final [ ]'s expressions from above.

$$[CH_3COOH] = 0.125 - 1.5 \times 10^{-3} M \approx 0.124 M$$

$$[CH_3COO^-] = [H^+] = 1.5 \times 10^{-3} M$$

Use the calculated value of [H$^+$] for the pH.

$$pH = -\log[H^+] = -\log(1.5 \times 10^{-3}) = 2.82$$

$$\% \text{ ionization} = \frac{[\text{ionized acid}]}{[\text{initial acid}]} \times 100 = \frac{1.5 \times 10^{-3}}{0.125} \times 100 = 1.2\%$$

A pH of 2.82 is reasonable for acidic solutions.

[ionized acid] = [H$^+$]   [initial acid] = original acid concentration

Hydrolysis or Solvolysis Solution

6. *What are the pH and pOH of an aqueous 0.125 M sodium acetate, NaCH$_3$COO, solution?* K$_a$ = 1.8 x 10^{-5}
   The correct answer is: pH = 8.92 and pOH =5.08.

INSIGHT:

a) The chemical compound ionizing is a soluble salt of a weak acid or base. ($NaCH_3COO$ is the soluble salt of acetic acid.)

b) A K_a or K_b will be provided but the salt contains the conjugate base or conjugate acid. You will have to use $K_w = K_a \times K_b$ to get the required ionization constant.

c) **If the salt contains the anion of a weak acid, the solution will be basic.** You will need the K_b for the ionization calculation.

d) **If the salt contains the cation of a weak base, the solution will be acidic.** You will need the K_a for the ionization calculation.

e) The ionization equation will involve the reaction of the salt with water. One of the ions will be a spectator ion which can be ignored in the ionization equation. (In this exercise, Na^+ is the spectator ion.)

f) Questions will commonly ask for pH or pOH of the solution.

g) The simplifying assumption used in exercise 4 usually will be applicable in these problems as the ionized concentrations are small.

$$CH_3COO^-(aq) + H_2O \rightleftharpoons CH_3COOH(aq) + OH^-(aq)$$

Starting [ ]'s	**0.125 M**	**0.00 M**	**0.00 M**
Change in [ ]'s	**- x**	**+ x**	**+ x**
Final [ ]'s	**0.125 - x M**	**+ x M**	**+ x M**

Acetate ion, a good base, reacts with water to produce hydroxide ions.

Use these algebraic quantities in the K_b expression.

$$K_w = K_a \times K_b \text{ thus } \frac{K_w}{K_a} = K_b$$

$$K_b = \frac{1.00 \times 10^{-14}}{1.8 \times 10^{-5}} = 5.6 \times 10^{-10}$$

$$K_b = \frac{[CH_3COO^-][OH^-]}{[CH_3COOH]} = 5.6 \times 10^{-10}$$

CH_3COO^- ion is the conjugate base of acetic acid. We must have the K_b, not the K_a, to work this problem.

<table>
<tr><td>

These are the algebraic quantities from above.

</td><td>

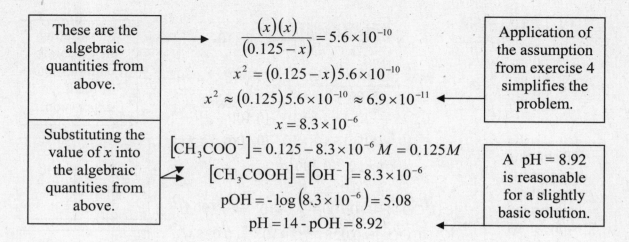

$$\frac{(x)(x)}{(0.125 - x)} = 5.6 \times 10^{-10}$$

$$x^2 = (0.125 - x)5.6 \times 10^{-10}$$

$$x^2 \approx (0.125)5.6 \times 10^{-10} \approx 6.9 \times 10^{-11}$$

</td><td>

Application of the assumption from exercise 4 simplifies the problem.

</td></tr>
</table>

$$x = 8.3 \times 10^{-6}$$

$$[CH_3COO^-] = 0.125 - 8.3 \times 10^{-6}\,M = 0.125M$$

$$[CH_3COOH] = [OH^-] = 8.3 \times 10^{-6}$$

$$pOH = -\log(8.3 \times 10^{-6}) = 5.08$$

$$pH = 14 - pOH = 8.92$$

Substituting the value of x into the algebraic quantities from above.

A pH = 8.92 is reasonable for a slightly basic solution.

Buffer Solution

7. ***What are the concentrations of the relevant species and the pH of a solution that is 0.100 M in acetic acid, CH₃COOH, and 0.025 M in sodium acetate, NaCH₃COO? $K_a = 1.8 \times 10^{-5}$***

   The correct answer is: $[CH_3COOH] = 0.100\,M$, $[CH_3COO^-] = 0.025\,M$, $[H^+] = 7.2 \times 10^{-5}\,M$, and pH = 4.14.

INSIGHT:

Key clues to recognizing a buffer problem:

a) The solution will contain a soluble salt dissolved in either a weak acid or a weak base. Concentrations or amounts of both will be present in the problem.

b) The salt must be the conjugate partner of the weak acid or weak base.

c) The equilibrium table will have starting concentrations of both the salt and the acid or base. The equilibrium will involve the common ion effect.

d) Henderson-Hasselbalch equations are a simple method to find the buffer solution's pH.

e) Simplifying assumption will be useful to quickly solve the solution concentration problem.

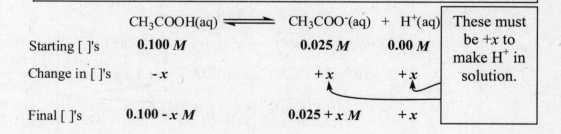

	CH₃COOH(aq) ⇌	CH₃COO⁻(aq) +	H⁺(aq)	
Starting [ ]'s	**0.100 *M***	**0.025 *M***	**0.00 *M***	These must be +x to make H⁺ in solution.
Change in [ ]'s	**-*x***	**+*x***	**+*x***	
Final [ ]'s	**0.100 - *x* *M***	**0.025 + *x* *M***	**+*x***	

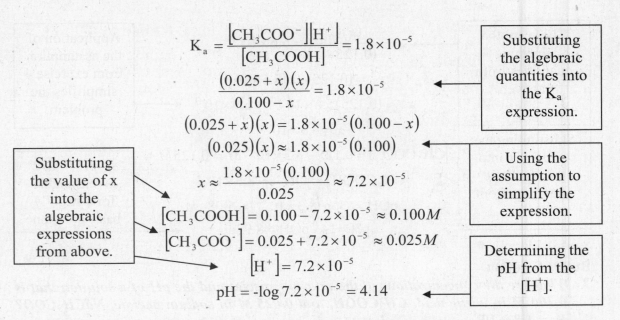

$$K_a = \frac{\left[CH_3COO^-\right]\left[H^+\right]}{\left[CH_3COOH\right]} = 1.8 \times 10^{-5}$$

$$\frac{(0.025 + x)(x)}{0.100 - x} = 1.8 \times 10^{-5}$$

$$(0.025 + x)(x) = 1.8 \times 10^{-5}(0.100 - x)$$

$$(0.025)(x) \approx 1.8 \times 10^{-5}(0.100)$$

$$x \approx \frac{1.8 \times 10^{-5}(0.100)}{0.025} \approx 7.2 \times 10^{-5}$$

$$\left[CH_3COOH\right] = 0.100 - 7.2 \times 10^{-5} \approx 0.100 M$$

$$\left[CH_3COO^-\right] = 0.025 + 7.2 \times 10^{-5} \approx 0.025 M$$

$$\left[H^+\right] = 7.2 \times 10^{-5}$$

$$pH = -\log 7.2 \times 10^{-5} = 4.14$$

Substituting the algebraic quantities into the K_a expression.

Using the assumption to simplify the expression.

Substituting the value of x into the algebraic expressions from above.

Determining the pH from the $[H^+]$.

If the question only asks for the pH of the buffer solution, the simplest method to get that answer is to use the Henderson-Hasselbalch equations.

Solubility Product

8. The solubility of iron(II) hydroxide, Fe(OH)₂, in water is 1.1 x 10⁻³ g/L at 25.0°C. What is the solubility product constant for Fe(OH)₂?
The correct answer is: K_sp = 6.9 x 10⁻¹⁵.

INSIGHT: Key clues to recognizing solubility product problems:
a) The problem will have an insoluble salt dissolved in water.
b) Usually a K_{sp} value will be given or asked for.
c) In calculating the K_{sp} value, remember to include the stoichiometric factors in both the equilibrium and the K_{sp} calculations.
d)   The concentrations of the ions and the K_{sp} values will be quite small.

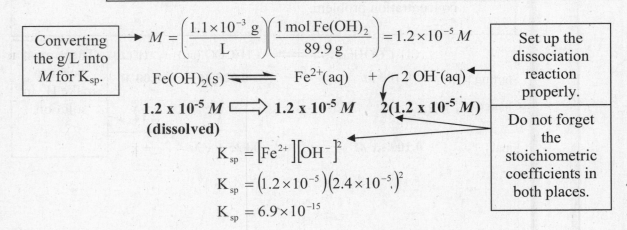

Converting the g/L into M for K_{sp}.

$$M = \left(\frac{1.1 \times 10^{-3} \text{ g}}{L}\right)\left(\frac{1 \text{ mol Fe(OH)}_2}{89.9 \text{ g}}\right) = 1.2 \times 10^{-5} M$$

$$Fe(OH)_2(s) \rightleftharpoons Fe^{2+}(aq) + 2 OH^-(aq)$$

Set up the dissociation reaction properly.

1.2 x 10⁻⁵ M ⟹ 1.2 x 10⁻⁵ M 2(1.2 x 10⁻⁵ M)
(dissolved)

Do not forget the stoichiometric coefficients in both places.

$$K_{sp} = \left[Fe^{2+}\right]\left[OH^-\right]^2$$

$$K_{sp} = \left(1.2 \times 10^{-5}\right)\left(2.4 \times 10^{-5}\right)^2$$

$$K_{sp} = 6.9 \times 10^{-15}$$

9. What are the molar solubilities of Zn^{2+} and OH^- for zinc hydroxide at 25.0°C? $K_{sp} = 4.5 \times 10^{-17}$

The correct answer is: $[Zn^{2+}] = 2.2 \times 10^{-5}$ M, $[OH^-] = 4.4 \times 10^{-5}$ M.

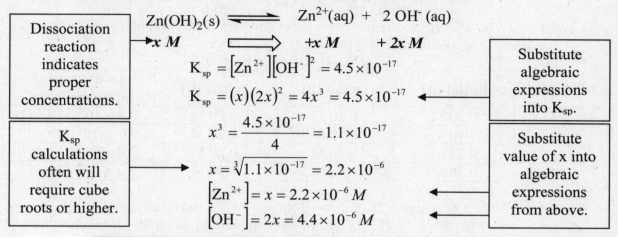

Dissociation reaction indicates proper concentrations.

$Zn(OH)_2(s) \rightleftharpoons Zn^{2+}(aq) + 2 OH^- (aq)$

$x M \implies +x M \qquad + 2x M$

$K_{sp} = [Zn^{2+}][OH^-]^2 = 4.5 \times 10^{-17}$

$K_{sp} = (x)(2x)^2 = 4x^3 = 4.5 \times 10^{-17}$

Substitute algebraic expressions into K_{sp}.

$x^3 = \dfrac{4.5 \times 10^{-17}}{4} = 1.1 \times 10^{-17}$

$x = \sqrt[3]{1.1 \times 10^{-17}} = 2.2 \times 10^{-6}$

$[Zn^{2+}] = x = 2.2 \times 10^{-6} M$

$[OH^-] = 2x = 4.4 \times 10^{-6} M$

K_{sp} calculations often will require cube roots or higher.

Substitute value of x into algebraic expressions from above.

Tips on aqueous equilibrium problems:
a) The hardest part is deciding if the problem is a weak acid/base, a solvolysis, a buffer, or a solubility product problem. The **INSIGHT** columns provide the key clues in determining which problem you are working on. Be very familiar with these.
b) Always write the correct ionization reaction and ionization expression (K_a, K_b, K_{sp}, etc.) for the problem you are given then set up the appropriate table underneath the ionization reaction to help with the algebra.
c) Correct use of the simplifying assumption will save enormous amounts of time and yield the correct answer. Use it whenever you can!
d) Every one of these equilibrium problems uses the same basic mathematical method to determine the concentrations and pH, etc. The only differences are setting up the ionization reactions and placing concentrations in the appropriate places in the table. If you can recognize the problem type, solving the problem is very straightforward after that.
e) The hardest equilibrium for students to recognize is always the hydrolysis problems. Keep an eye out for these.

Practice Test Five
Modules 16-18

Level 1   1. Determine the rate-law expression for the reaction below using the experimental data provided.

$$2A + B + C \rightarrow D + E$$

Trial	Initial [A]	Initial [B]	Initial [C]	Initial rate of formation of BC
1	0.20 M	0.20 M	0.20 M	2.4×10^{-6} M·min^{-1}
2	0.40 M	0.30 M	0.20 M	9.6×10^{-6} M·min^{-1}
3	0.20 M	0.30 M	0.20 M	2.4×10^{-6} M·min^{-1}
4	0.20 M	0.40 M	0.60 M	7.2×10^{-6} M·min^{-1}

Level 1   2. The reaction $2N_2O_{5(g)} \rightarrow 2N_2O_{4(g)} + O_{2(g)}$ has a rate constant of $0.00840 s^{-1}$. If 2.25 mol of N_2O_5 are placed in a 4.00 L container, then what would be the concentration of N_2O_5 after 2.50 minutes? Is the reaction first order or second order with respect to N_2O_5?

Level 1   3. The decomposition of $NOBr_{(g)}$ to $NO_{(g)}$ and $Br_{2(g)}$ has $k = 0.810 \ M^{-1}s^{-1}$. If the initial concentration of NOBr is $4.00 \times 10^{-3} \ M$, then how long does it take for the concentration of NOBr to decrease to $1.50 \times 10^{-4} \ M$?

Level 1   4. Calculate E_a for a reaction in which the rate is $1.2 \times 10^2 \ s^{-1}$ at 273 K and $3.6 \times 10^2 \ s^{-1}$ at 298 K.

Level 1   5. Calculate K_c for the following reaction given that the equilibrium concentrations are: $[F_2] = 1.8 \times 10^{-3} \ M$; $[Br_2] = 9.0 \times 10^{-3} \ M$; $[BrF_5] = 4.6 \times 10^{-3} \ M$. $2BrF_{5(g)} \leftrightarrow Br_{2(g)} + 5F_{2(g)}$

Level 3   6. At a given temperature, K_c for the reaction below is 0.0104. If the initial concentrations of $PCl_{5(g)}$ is 0.55 M then what is the equilibrium concentration of PCl_3?

$$PCl_5 \rightarrow PCl_{3(g)} + Cl_{2(g)}$$

Level ?   7. If a reaction is endothermic, then will increasing the temperature at which the reaction occurs favor the products or the reactants? Explain.

Level 1   8. What is the pH of a solution in which the hydroxide ion concentration is 1.83×10^{-7}?

Level 3   9. Determine the pH and the pOH of a 0.025 M solution of potassium nitrite. The K_a of HNO_2 is 4.5×10^{-4}.

Level 3   10. What is the pH of a solution in which the ratio of benzoic acid to sodium benzoate is 2:1. The K_a of benzoic acid is 6.3×10^{-5}.

Module 19 Predictor Questions

The following questions may help you to determine to what extent you need to study this module. The questions are ranked according to ability.

Level 1 = basic proficiency
Level 2 = mid level proficiency
Level 3 = high proficiency

If you can correctly answer the Level 3 questions, then you probably do not need to spend much time with this module. If you are only able to answer the Level 1 problems, then you should review the topics covered in this module.

Level 3   1. Balance the following oxidation-reduction reaction in acidic solution. Identify the species that is oxidized, the species that is reduced, the oxidizing agent, and the reducing agent.

$$I_2 (s) + S_2O_3^{2-}(aq) \rightarrow I^- (s) + S_4O_6^{2-}(aq)$$

Level 3   2. Balance the following oxidation-reduction reaction in acidic solution. Identify the species that is oxidized, the species that is reduced, the oxidizing agent, and the reducing agent.

$$CrO_2^- + ClO^- \rightarrow CrO_4^{2-} + Cl^-$$

Level 2   3. What mass of iron metal is produced at the cathode when 2.50 amps of current are passed through an electrolytic cell containing iron (II) nitrate for 65 minutes?

Level 2   4. A voltaic cell contains a 1.0 M solution of $Ni(NO_3)_2$ and a 1.0 M solution of $CuSO_4$. Draw the cell and determine what chemical species will be produced at the cathode and at the anode. What is the direction of electron flow in the cell?

Level 1   5. Determine the standard cell potential for the voltaic cell in question number 4.

Level 2   6. What is the cell potential at 338 K for the voltaic cell in question number 4 if the $Ni(NO_3)_2$ concentration is 2.5 M and the $CuSO_4$ concentration is 1.3 M?

Level 1   7. Determine the equilibrium constant for the voltaic cell in question number 4 at 298 K.

Module 19 Predictor Question Solutions

1. $I_2 + S_2O_3{}^{2-} \rightarrow I^- + S_4O_6{}^{2-}$

   $I_2 + 2e^- \rightarrow 2I^-$ reduction half-reaction
   $2 S_2O_3{}^{2-} \rightarrow S_4O_6{}^{2-} + 2e^-$ oxidation half-reaction

   $I_2 + 2S_2O_3{}^{2-} \rightarrow 2I^- + S_4O_6{}^{2-}$ complete balanced reaction

   I_2 is reduced, and it is the oxidizing agent.
   $S_2O_3{}^{2-}$ is oxidized, and it is the reducing agent.

2. $CrO_2{}^- + ClO^- \rightarrow CrO_4{}^{2-} + Cl^-$

   $(ClO^- + 2 H^+ + 2e^- \rightarrow Cl^- + H_2O)3$ reduction half-reaction
   $(CrO_2{}^- + 2 H_2O \rightarrow CrO_4{}^{2-} + 4 H^+ + 3e^-)2$ oxidation half-reaction

   $2 CrO_2{}^- + 4 H_2O + 3ClO^- + 6 H^+ \rightarrow 2 CrO_4{}^{2-} + 8 H^+ + 3Cl^- + 3H_2O$ complete reaction

   Since the reaction takes place in basic solution, H^+ cannot exist. It will combine with OH^- to form water. Eight OH^- must be added to each side of the reaction, and the resulting balanced equation is:
   $2 CrO_2{}^- + 3ClO^- + 2 OH^- \rightarrow 2 CrO_4{}^{2-} + 3Cl^- + H_2O$

3.

   $$2.50 \text{ amps} = \frac{2.50 \text{ C}}{\text{s}}$$

   $$(65 \text{ min})\left(\frac{60 \text{ s}}{1 \text{ min}}\right) = (3900 \text{ s})$$

   $$(3900 \text{ s})\left(\frac{2.50 \text{ C}}{\text{s}}\right) = 9750 \text{ C}$$

   $$(9750 \text{ C})\left(\frac{1 \text{ mol e}^-}{9.65 \times 10^4 \text{ C}}\right)\left(\frac{1 \text{ mol Fe}}{2 \text{ mol e}^-}\right)\left(\frac{55.85 \text{ g Fe}}{1 \text{ mol Fe}}\right) = 2.82 \text{ g Fe}_{(s)}$$

4. In voltaic cells, the oxidation reaction has the more negative standard reduction potential, and the reduction reaction has the more positive standard reduction potential.

   $Ni^{2+} + 2e^- \rightarrow Ni_{(s)}$ $E^0 = -0.25V$ more negative = oxidation
   $Cu^{2+} + 2e^- \rightarrow Cu_{(s)}$ $E^0 = 0.337V$ more positive = reduction

   $Ni_{(s)}$ is oxidized to Ni^{2+} at the anode, and Cu^{2+} is reduced to $Cu_{(s)}$ at the cathode. Electron flow is from the anode to the cathode (as in all voltaic cells).

5. The sign of E^0 for the oxidation half-reaction must be reversed, along with the reaction.

$$Ni_{(s)} \rightarrow Ni^{2+} + 2e^- \qquad\qquad E^0 = +0.250V$$
$$\underline{Cu^{2+} + 2e^- \rightarrow Cu_{(s)} \qquad\qquad E^0 = 0.337V}$$
$$Ni_{(s)} + Cu^{2+} \rightarrow Ni^{2+} + Cu_{(s)} \qquad E^0{}_{cell} = 0.587V$$

6.

$$E = E^0 - \left(\frac{2.303RT}{nF}\right)\left(\log\frac{[\text{oxidized species}]}{[\text{reduced species}]}\right)$$

$$E = 0.587V - \left(\frac{(2.303)(8.314\frac{J}{mol\cdot K})(338\,K)}{(2\,mol\,e^-)(9.65 \times 10^4\frac{J}{V\cdot mol\,e^-})}\right)\left(\log\frac{[2.5]}{[1.3]}\right) = 0.577\ V$$

7.

$$nFE^0{}_{cell} = RT\ln K \Rightarrow \ln K = \frac{nFE^0{}_{cell}}{RT}$$

$$\ln K = \frac{(2\,mol\,e^-)(9.65\times 10^4\frac{J}{V\cdot mol\,e^-})(0.587\,V)}{(8.314\frac{J}{mol\cdot K})(298\,K)} = 45.73$$

$$K = 7.22 \times 10^{19}$$

Module 19
Electrochemistry

Introduction

This module describes the basic electrochemistry methods used in typical textbooks. The goals of the module are to describe how to:

1. balance redox reactions in acidic and basic solutions
2. discern between electrolytic and voltaic cells
3. use Faraday's law to determine the amount of a species that is reduced in an electrolytic cell
4. determine the anode, cathode, and electron flow in both cell types
5. determine the standard cell potential for a voltaic cell
6. use the Nernst equation to find the cell potential at nonstandard conditions
7. determine the Gibb's Free Energy change and equilibrium constant for a cell from its standard potential

You will need your textbook opened to the appendix containing the standard electrode potentials to understand this material.

Module 19 Key Equations & Concepts

1. **Electrolytic cells are electrochemical cells in which nonspontaneous chemical reactions are forced to occur by the application of an external voltage.**

   In electrolytic cells reactions that would not occur in nature, such as the electrolysis of chemical compounds or the electroplating of metals, are made to occur by the passage of electricity through the cell.

2. **Voltaic cells are electrolytic cells in which spontaneous chemical reactions occur and the electrons generated in the reaction are passed through an external wire.**

   Voltaic cells are batteries such as dry cells and lithium batteries used in watches, cameras, etc.

3. **The anode is the electrode where oxidation occurs in both electrolytic and voltaic cells. The cathode is the electrode where reduction occurs in both cell types.**

   In electrolytic cells the anode is the positive electrode and the cathode is the negative electrode. This is reversed for voltaic cells where the anode is the negative electrode and the cathode is the positive electrode.

4. **Faraday's law of electrolysis states that the amount of a chemical compound oxidized or reduced at an electrode during electrolysis is directly proportional to the amount of electricity passed through the cell.**

   This law is used to calculate the number of grams of a chemical compound transformed from oxidized to reduced species, or vice versa, in an electrolytic cell. Faraday's constant, 1 faraday = 9.65×10^4 coulombs, is essential in these calculations.

5. **Standard Cell Potentials are the initial voltage produced in a voltaic cell at standard conditions.**

    To find the standard cell potential, add the standard cell potential for the reduction step to the reverse of the standard cell potential for the oxidation step. Standard cell potentials are tabulated in your textbook in the appendices.

6. $$E = E^0 - \frac{2.303\ RT}{n\ F} \log Q \text{ or } E = E^0 - \frac{0.0592}{n} \log Q \text{ at } 25.0^\circ C$$

    The Nernst equation is used to calculate a cell's potential at nonstandard conditions. E is the nonstandard cell potential, E^0 is the standard cell potential, n is the number of moles of electrons in the reaction, F is Faraday's constant, and Q is the reaction quotient.

7. $\Delta G^0 = -nFE^0_{cell}$ **or** $nFE^0_{cell} = RT \ln K$

    This equation is used to determine either the Gibbs Free Energy change or the equilibrium constant of a chemical reaction once the cell potential has been determined.

Sample Exercises
Balancing Redox Reactions
1. Balance the following redox reaction in acidic solution.

$$Cu(s) + NO_3^-(aq) \rightarrow Cu^{2+}(aq) + NO_2(g)$$

The correct answer is:

$$Cu(s) + 4\ H^+(aq) + 2\ NO_3^-(aq) \rightarrow Cu^{2+}(aq) + 2\ NO_2(g) + 2\ H_2O(\ell).$$

INSIGHT: Redox reactions can occur in either acidic or basic solutions. In acidic solutions you can add H^+ and H_2O to balance the reaction. In basic solutions you can add OH^- and H_2O. The problem will either state that the reaction is in acidic solution, or H^+ will be present as a reactant or product. Similarly, for basic solutions look for a statement that the reaction is in basic solution or the presence of OH^-.

 YIELD

There are two simple methods to balance redox reactions, the change in oxidation number method and the half-reaction method. The change in oxidation number method is more physically correct but the half-reaction method is simpler to learn and more straight forward. We will use the half-reaction method in these exercises. It is very important that you review the rules for assigning oxidation numbers and for balancing redox reactions found in your textbook.

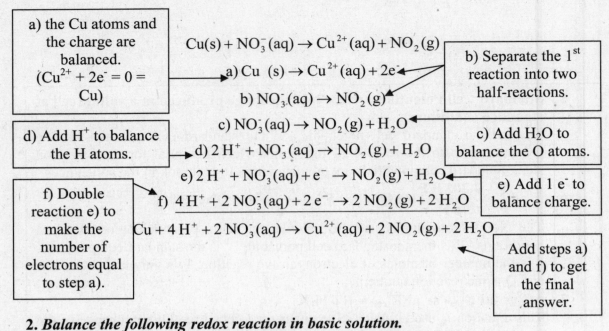

a) the Cu atoms and the charge are balanced. ($Cu^{2+} + 2e^- = 0 = Cu$)

d) Add H^+ to balance the H atoms.

f) Double reaction e) to make the number of electrons equal to step a).

$Cu(s) + NO_3^-(aq) \rightarrow Cu^{2+}(aq) + NO_2(g)$

a) $Cu\ (s) \rightarrow Cu^{2+}(aq) + 2e^-$

b) $NO_3^-(aq) \rightarrow NO_2(g)$

c) $NO_3^-(aq) \rightarrow NO_2(g) + H_2O$

d) $2\,H^+ + NO_3^-(aq) \rightarrow NO_2(g) + H_2O$

e) $2\,H^+ + NO_3^-(aq) + e^- \rightarrow NO_2(g) + H_2O$

f) $4\,H^+ + 2\,NO_3^-(aq) + 2\,e^- \rightarrow 2\,NO_2(g) + 2\,H_2O$

$Cu + 4\,H^+ + 2\,NO_3^-(aq) \rightarrow Cu^{2+}(aq) + 2\,NO_2(g) + 2\,H_2O$

b) Separate the 1st reaction into two half-reactions.

c) Add H_2O to balance the O atoms.

e) Add 1 e^- to balance charge.

Add steps a) and f) to get the final answer.

2. Balance the following redox reaction in basic solution.
$$CrO_2^-(aq) + ClO^-(aq) \rightarrow CrO_4^-(aq) + Cl^-(aq)$$

The correct answer is: $CrO_2^-(aq) + 2ClO^-(aq) \rightarrow CrO_4^-(aq) + 2Cl^-(aq)$.

a) Separate the reaction into the reduction and oxidation half reactions.

e) Balance H and O atoms in half-reaction d) by adding OH^- and H_2O.

i) Add steps g) and h).

$CrO_2^- + ClO^- \rightarrow CrO_4^- + Cl^-$

a) $CrO_2^- \rightarrow CrO_4^-$

b) $CrO_2^- + 4OH^- \rightarrow CrO_4^- + 2H_2O$

c) $CrO_2^- + 4OH^- \rightarrow CrO_4^- + 2H_2O + 4e^-$

d) $ClO^- \rightarrow Cl^-$

e) $ClO^- + H_2O \rightarrow Cl^- + 2OH^-$

f) $ClO^- + H_2O + 2e^- \rightarrow Cl^- + 2OH^-$

g) $CrO_2^- + 4OH^- \rightarrow CrO_4^- + 2H_2O$

h) $2ClO^- + 2H_2O \rightarrow 2Cl^- + 4OH^-$

i) $CrO_2^- + 4OH^- + 2ClO^- + 2H_2O \rightarrow CrO_4^- + 2H_2O + 2Cl^- + 4OH^-$

j) $CrO_2^- + 2ClO^- \rightarrow CrO_4^- + 2Cl^-$

b) Balance H and O atoms in half-reaction a) by adding OH^- and H_2O.

c) Add 4 e^- to step b) to balance charge.

f) Add 2 e^- to step e) to balance charge.

g) The equations from step c) and f) (2x) are added.

j) Remove 4 OH^- and 2 H_2O from both sides.

Electrolytic Cell

3. *An electrolytic cell containing an aqueous NaCl solution is constructed. What chemical species will be produced at the cathode and anode? What is the direction of electron flow in this cell?*

The correct answer is: $H_2(g)$ is produced at the cathode and $Cl_2(g)$ is produced at the anode. The electrons flow from the anode passing through the battery and to the cathode.

INSIGHT: In this electrolytic cell there are four possible redox reactions. The two possible reductions are $Na^+(aq)$ to $Na(s)$ or H_2O to $H_2(g)$ and $OH^-(aq)$. The two possible oxidations are $Cl^-(aq)$ to $Cl_2(g)$ or H_2O to $H^+(aq)$ and O_2. *How do you determine the correct reactions in electrolytic cells?* Electrolytic cells force nonspontaneous chemical reactions to occur. The standard reduction potentials in your textbook will tell you which one to choose. 1) The oxidation reaction will be the one of the two possible reactions that has the most positive reduction potential (most negative oxidation potential). 2) The reduction reaction will be the one of the two possible reactions that has the most positive, or least negative, reduction potential.

In electrolytic cells, the - electrode is the cathode (reduction), the + electrode is the anode (oxidation).

In electrolytic cells, the electrons flow from the anode, + electrode, to the cathode, - electrode.

Reduction Potentials
$$2H_2O+2e^-\rightarrow H_2+4H^+ +4e^-$$
$$E^0_{cell} = -0.828 \text{ V}$$
$$Na^+ + e^- \rightarrow Na$$
$$E^0_{cell} = -2.71 \text{ V}$$
-0.828 is more positive than -2.71 so the 1st reaction occurs.

Oxidation Potentials
$$2Cl^-\rightarrow Cl_2 + 2e^-$$
$$E^0_{cell} = -1.36 \text{ V}$$
$$2H_2O\rightarrow O_2+4H^+ +4e^-$$
$$E^0_{cell} = -1.23 \text{ V}$$
-1.36 is more negative than -1.23 so the 1st reaction occurs.

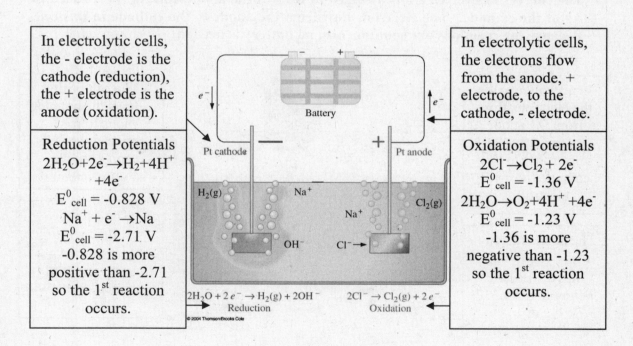

Faraday's Law

4. *How many grams of nickel metal will be produced at the cathode when 3.75 amps of current are passed for 75.0 minutes through an electrolytic cell containing $NiSO_{4(aq)}$?*

The correct answer is: 5.15 g of Ni.

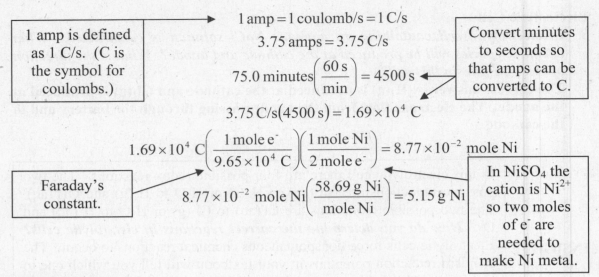

1 amp is defined as 1 C/s. (C is the symbol for coulombs.)	$1\,amp = 1\,coulomb/s = 1\,C/s$ $3.75\,amps = 3.75\,C/s$ $75.0\,minutes\left(\dfrac{60\,s}{min}\right) = 4500\,s$ $3.75\,C/s(4500\,s) = 1.69 \times 10^4\,C$	Convert minutes to seconds so that amps can be converted to C.

$$1.69 \times 10^4\,C\left(\frac{1\,mole\,e^-}{9.65 \times 10^4\,C}\right)\left(\frac{1\,mole\,Ni}{2\,mole\,e^-}\right) = 8.77 \times 10^{-2}\,mole\,Ni$$

Faraday's constant.	$8.77 \times 10^{-2}\,mole\,Ni\left(\dfrac{58.69\,g\,Ni}{mole\,Ni}\right) = 5.15\,g\,Ni$

In $NiSO_4$ the cation is Ni^{2+} so two moles of e^- are needed to make Ni metal.

Voltaic Cell

5. A voltaic cell is constructed of a 1.0 M CuSO₄(aq) solution and a 1.0 M AgNO₃(aq) solution plus the electrodes, connecting wires, and salt bridges. What chemical species will be produced at the cathode and anode? What is the direction of electron flow in this cell?

The correct answer is: Cu is oxidized to Cu^{2+} at the anode and Ag^+ is reduced to Ag at the cathode. The electrons flow from the anode to the cathode in this cell. Because the reactions are spontaneous, no battery is needed.

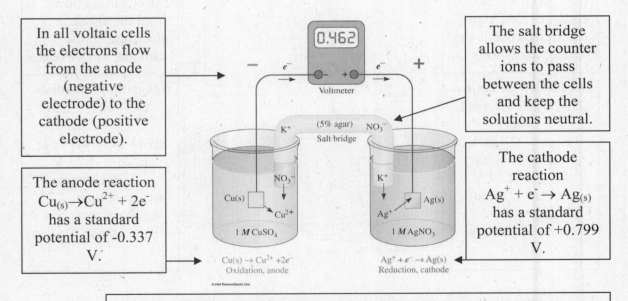

In all voltaic cells the electrons flow from the anode (negative electrode) to the cathode (positive electrode).

The salt bridge allows the counter ions to pass between the cells and keep the solutions neutral.

The anode reaction $Cu_{(s)} \rightarrow Cu^{2+} + 2e^-$ has a standard potential of -0.337 V.

The cathode reaction $Ag^+ + e^- \rightarrow Ag_{(s)}$ has a standard potential of +0.799 V.

INSIGHT: In voltaic cells the reactions are spontaneous. You can predict the reactions using the standard reduction potentials. 1) The oxidation reaction will have the least positive (most negative) standard reduction potential. 2) The reduction reaction will have the most positive (least negative) standard reduction potential.

186

6. What is the standard cell potential for the voltaic cell in Sample Exercise 5?
    The correct answer is: 0.462 V.

| The reduction reaction is balanced **but the E^0_{cell} is not doubled**. Cell potentials are intensive quantities. | $$2\,Ag^+(aq) + 2\,e^- \rightarrow 2\,Ag(s) \qquad E^0 = +0.799\,V$$ $$Cu(s) \rightarrow Cu^{2+}(aq) + 2\,e^- \qquad E^0 = -0.337\,V$$ <hr> $$2\,Ag^+(aq) + Cu(s) \rightarrow 2\,Ag_{(s)} + Cu^{2+}(aq) \quad E^0_{cell} = +0.462\,V$$ The cell potentials, E^0, are from the tables in the appendix of your textbook. In voltaic cells, the overall cell potential E^0_{cell} must be a <u>positive</u> value. | Always reverse the reaction and change the sign of the tabulated reduction potential for the oxidation. |

YIELD

To calculate the standard cell potential follow these steps:
1) Write the half-reaction and the cell potential for the reaction that has the most positive (or least negative) reduction potential, E^0.
2) Write the half-reaction and the cell potential for the other reaction as an oxidation. To do this you must take the reaction given in the table in your textbook, reverse the reaction and change the sign of E^0.
3) Make sure that the electrons from each half-reaction are balanced but **<u>do not multiply the reduction potentials</u>**.
4) Add the two half-reactions, canceling the electrons, and add the two cell potentials to get the E^0_{cell}. This must always be a positive value to indicate that the reaction is spontaneous.

Nernst Equation
7. What is the cell potential for the voltaic cell in Sample Exercise 5 at 325 K if the $CuSO_4$ concentration is 2.00 M and the $AgNO_3$ concentration is 3.00 M?

   The correct answer is: 0.483 V.

| E^0 is the cell potential calculated in exercise 4. R is the gas constant. n is the moles of electrons in the reaction. | $$E = E^0 - \frac{2.303\,RT}{n\text{F}}\log Q$$ $$E = 0.462\,V - \frac{2.303\,(8.314\,J/mol\,K)(325\,K)}{(2\,mol)(9.65 \times 10^4\,J/V\,mol\,e^-)}\log\frac{2.00}{(3.00)^2}$$ $$E = 0.462\,V - \frac{6223\,V}{1.93 \times 10^5}\log\frac{2}{9}$$ $$E = 0.462\,V - \frac{6223\,V}{1.93 \times 10^5}(-0.653)$$ $$E = 0.462\,V - (-0.021\,V) = 0.483\,V$$ | F is faraday's constant. Q is the reaction quotient as described in module 17. |

YIELD	The cell potential determined in exercise 3 is at standard conditions (298 K, 1 *M* solutions, 1 atm pressure, etc.). The Nernst equation allows us to calculate the cell potential at nonstandard conditions such as different temperatures and solution concentrations.

INSIGHT:	1) Because 1 J = V·C thus $(9.65 \times 10^4$ C/mol $e^-)(1$ J/V·C$) = 9.65 \times 10^4$ J/V mol e^-, which is Faraday's constant in a different set of units used in this exercise. 2) This is a heterogeneous equilibrium, there are solids and solutions involved in the reaction. Equilibrium constants, including Q the reaction quotient, for heterogeneous equilibria involve only the species having the largest activities. Solutions have much larger activities than solids and therefore we only use the solution concentrations in Q. 3) The cell reaction is 2 Ag^+(aq) + Cu(s) → 2 Ag(s) + Cu^{2+}(aq). Thus we get $Q = \dfrac{\left[Cu^{2+}\right]}{\left[Ag^+\right]^2} = \dfrac{2.00}{(3.00)^2} = \dfrac{2}{9}$.

Determination of the Equilibrium Constant for a Voltaic Cell

8. *What is the equilibrium constant at 298 K for the voltaic cell described in Sample Exercise 5?*

The correct answer is: 4.31 x 10¹⁵.

The correct answer is: 4.31×10^{15}.

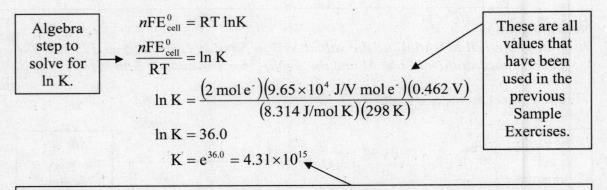

Algebra step to solve for ln K.	$nFE^0_{cell} = RT \ln K$ $\dfrac{nFE^0_{cell}}{RT} = \ln K$ $\ln K = \dfrac{(2 \text{ mol } e^-)(9.65 \times 10^4 \text{ J/V mol } e^-)(0.462 \text{ V})}{(8.314 \text{ J/mol K})(298 \text{ K})}$ $\ln K = 36.0$ $K = e^{36.0} = 4.31 \times 10^{15}$	These are all values that have been used in the previous Sample Exercises.

This cell reaction is spontaneous and also very product favored. The very large value of K indicates that the reaction should give large amounts of products.

Module 20 Predictor Questions

The following questions may help you to determine to what extent you need to study this module. The questions are ranked according to ability.

> Level 1 = basic proficiency
> Level 2 = mid level proficiency
> Level 3 = high proficiency

If you can correctly answer the Level 3 questions, then you probably do not need to spend much time with this module. If you are only able to answer the Level 1 problems, then you should review the topics covered in this module.

Level 1 1. Calculate the mass defect, in amu, for a ^{45}Ca nucleus? The actual mass of a ^{45}Ca atom is 44.9562 amu.

Level 1 2. What is the binding energy, in J/mol, for a ^{45}Ca atom?

Level 1 3. What nuclide is the product of the alpha decay of ^{238}U?

Level 2 4. What nuclide is the product of the beta, β^-, decay of ^{137}Cs?

Level 2 5. What nuclide is the product of the positron, β^+, decay of ^{22}Na?

Level 1 6. Fill in the missing nuclide in the following nuclear reaction.

$$^{234}\text{Th} \rightarrow {}^{0}_{-1}\beta^- + \underline{\hspace{2cm}}$$

Level 1 7. Frequently used as a radioactive tracer in laboratory experiments, the nuclide ^{32}P has a half-life of 14.28 days. If 1.5 g of ^{32}P are used in an experiment, how much will be left 50 days later?

Level 2 8. The nuclide ^{14}C is frequently used to date artifacts. If a given artifact has a ^{14}C content of 20.2% and the half-life of ^{14}C is 5730 years, then how old is the artifact? Assume that the decrease in ^{14}C content is entirely due to radioactive decay.

Module 20 Predictor Question Solutions

1.

$$\Delta m = [Z(1.0073) + N(1.0087) + Z(0.0005)] - \text{actual mass}$$

For ^{45}Ca, $Z = 20$, $N = 25$, actual mass $= 44.9562$ amu

$$\Delta m = [20(1.0073) + 25(1.0087) + 20(0.0005)] - 44.9562 \text{ amu} = 0.4173 \text{ amu/atom}$$

$$\left(\frac{0.4173 \text{ amu}}{\text{atom}}\right)\left(\frac{1.000 \text{ g}}{6.022 \times 10^{23} \text{ amu}}\right)\left(\frac{60.22 \times 10^{23} \text{ atoms}}{1 \text{ mol}}\right) = 0.4173 \frac{\text{g}}{\text{mol atoms}}$$

2. $\Delta m = 0.4173$ g/mol atoms $= 4.173 \times 10^{-4}$ kg/mol atoms

   $BE = \Delta mc^2 = (4.173 \times 10^{-4} \text{ kg/mol atoms})(3.00 \times 10^8 \text{ m/s})^2 = 3.7557 \times 10^{13}$ J/mol atoms

3. $^{238}_{92}U \rightarrow {}^{4}_{2}He + \underline{{}^{234}_{90}Th}$

4. $^{22}_{11}Na \rightarrow {}^{0}_{+1}\beta + \underline{{}^{22}_{10}Ne}$

5. $^{137}_{55}Cs \rightarrow {}^{0}_{-1}\beta + \underline{{}^{137}_{56}Ba}$

6. $^{234}_{90}Th \rightarrow {}^{0}_{-1}\beta + \underline{{}^{234}_{91}Pa}$

7. $A = A_0 e^{-kt}$

   $$k = \frac{0.693}{t_{1/2}} = \frac{0.693}{14.28 \text{ d}} = 0.0485 \text{ d}^{-1}$$

   $$A = (1.5 \text{ g})e^{-(0.0485 \text{ d}^{-1})(50 \text{ d})} = 0.13 \text{ g}$$

8. $A = A_0 e^{-kt}$

   $$k = \frac{0.693}{t1/2} = \frac{0.693}{5730 \text{ y}} = 1.21 \times 10^{-4} \text{ y}^{-1}$$

   $$A = A_0 e^{-kt} \Rightarrow \ln\frac{A}{A_0} = -kt$$

   $$\ln\frac{20.2}{100.} = -(1.21 \times 10^{-4} \text{ y}^{-1})t$$

   $$t = 1.32 \times 10^4 \text{ y}$$

Module 20
Nuclear Chemistry

Introduction

This module discusses the basic relationships used in a typical nuclear chemistry chapter. The important topics described include:

1. calculating the mass defect and binding energy for a nucleus
2. predicting the products of alpha, negatron, and positron radioactive decays as well as of a nuclear reaction
3. problems associated with the kinetics of radioactive decay.

Module 20 Key Equations & Concepts

1. **$\Delta m = [Z(1.0073) + N(1.0087) + Z(0.0005)]$ – actual mass of atom**
    The mass defect of a nucleus is the difference in the sum of the masses of the protons, neutrons, and electrons in a nucleus minus the actual mass of the atom. This relationship describes how much of the nuclear mass has been converted into energy to bind the nucleus.

2. **Binding Energy $= \Delta mc^2$**
    The nuclear binding energy is the energy required to hold the protons and neutrons together in the nucleus. It is the mass defect converted from mass unit to energy units.

3. **$_Z^A X \rightarrow _{Z-2}^{A-4} Y + _2^4 He$**
    This is the basic equation for radioactive alpha decay. Alpha decay removes two protons and two neutrons, in the form of a ^4He nucleus, from the decaying nucleus converting the element X into a new element Y.

4. **$_Z^A X \rightarrow _{Z+1}^A Y + _{-1}^0 e \ (or \ _{-1}^0 \beta^-)$**
    Radioactive beta decay, β^- or negatron decay, converts a neutron into a proton by eliminating a high velocity electron, the β^- particle, from the nucleus. The decaying nucleus, X, is converted to a new nucleus, Y, having one additional proton and one less neutron.

5. **$_Z^A X \rightarrow _{Z-1}^A Y + _{+1}^0 e \ (or \ _{+1}^0 \beta^+)$**
    Radioactive positron decay, β^+, converts a proton into a neutron by eliminating a high velocity positive electron, the β^+ particle, from the nucleus. The decaying nucleus, X is converted to a new nucleus, Y, having one less proton and one more neutron.

6. **$_{Z_1}^{M_1} Q \rightarrow _{Z_2}^{M_2} R + _{Z_3}^{M_3} Y$ where $M_1 = M_2 + M_3$ and $Z_1 = Z_2 + Z_3$**
    This is the basic relationship for nuclear reactions, and radioactive decays. The proton numbers of the product nuclides (Z_2 and Z_3) must sum to the original nuclide's proton number, Z_1. The mass numbers of the product nuclides (M_2 and M_3) must also add up to the original nuclide's mass, M_1.

7. **$A = A_0 e^{-kt}$ and $k \ t_{1/2} = 0.693$**
    Radioactive decay obeys first order kinetics as described in Module 16. These are the two important equations for radioactive decay describing the amount of a nuclide remaining after a certain amount of time has passed and the half-life relationship.

Sample Exercises

Mass Defect

1. What is the mass defect, in amu, for a ^{55}Cr nucleus? The actual mass of a ^{55}Cr atom is 54.9408 amu.

The correct answer is: 0.5161 amu.

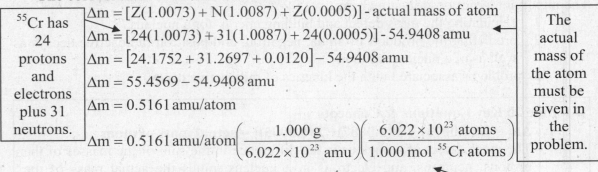

^{55}Cr has 24 protons and electrons plus 31 neutrons.

$\Delta m = [Z(1.0073) + N(1.0087) + Z(0.0005)]$ - actual mass of atom

$\Delta m = [24(1.0073) + 31(1.0087) + 24(0.0005)] - 54.9408$ amu

$\Delta m = [24.1752 + 31.2697 + 0.0120] - 54.9408$ amu

$\Delta m = 55.4569 - 54.9408$ amu

$\Delta m = 0.5161$ amu/atom

$\Delta m = 0.5161$ amu/atom $\left(\dfrac{1.000 \text{ g}}{6.022 \times 10^{23} \text{ amu}} \right) \left(\dfrac{6.022 \times 10^{23} \text{ atoms}}{1.000 \text{ mol } ^{55}Cr \text{ atoms}} \right)$

$\Delta m = 0.5161$ g/mol atoms $= 5.161 \times 10^{-4}$ kg/mol atoms

The actual mass of the atom must be given in the problem.

We will use the mass defect in kg/mol atoms in the next exercise.

There is 1.00 mole of amu in 1.000 g.

INSIGHT: 1.0073 is the mass of a proton, 1.0087 amu is the mass of a neutron, and 0.0005 amu is the mass of an electron.

Binding Energy

2. What is the binding energy, in J/mol, for a ^{55}Cr nucleus?

The correct answer is: 4.65 x 10^{13} J/mol atoms.

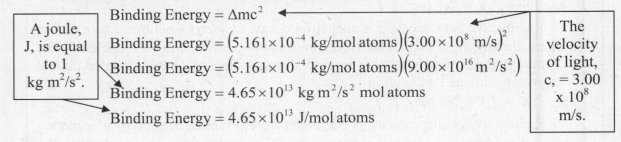

A joule, J, is equal to 1 kg m^2/s^2.

Binding Energy $= \Delta m c^2$

Binding Energy $= (5.161 \times 10^{-4} \text{ kg/mol atoms})(3.00 \times 10^{8} \text{ m/s})^2$

Binding Energy $= (5.161 \times 10^{-4} \text{ kg/mol atoms})(9.00 \times 10^{16} \text{ m}^2/\text{s}^2)$

Binding Energy $= 4.65 \times 10^{13}$ kg m^2/s^2 mol atoms

Binding Energy $= 4.65 \times 10^{13}$ J/mol atoms

The velocity of light, c, = 3.00 x 10^8 m/s.

Alpha Decay

3. What is the product nuclide of the alpha decay of ^{232}Th?

The correct answer is: ^{228}Ra.

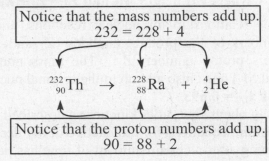

Notice that the mass numbers add up.
232 = 228 + 4

$^{232}_{90}Th \rightarrow {}^{228}_{88}Ra + {}^{4}_{2}He$

Notice that the proton numbers add up.
90 = 88 + 2

Beta Decay

4. What is the product nuclide of the β^-, negatron, decay of ^{14}C?

The correct answer is: ^{14}N.

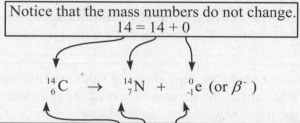

Notice that the mass numbers do not change.
$14 = 14 + 0$

$$^{14}_{6}C \rightarrow {}^{14}_{7}N + {}^{0}_{-1}e \text{ (or } \beta^-)$$

Notice that the charges of the protons and the beta particle add up.
$6 = 7 + (-1)$

5. What is the product nuclide of the β^+, positron, decay of ^{37}Ca?

The correct answer is: ^{37}K.

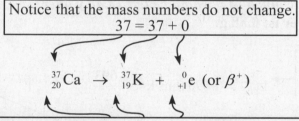

Notice that the mass numbers do not change.
$37 = 37 + 0$

$$^{37}_{20}Ca \rightarrow {}^{37}_{19}K + {}^{0}_{+1}e \text{ (or } \beta^+)$$

Notice that the charges of the protons and the beta particle add up.
$20 = 19 + (+1)$

Nuclear Reaction

6. Fill in the missing nuclide in this nuclear reaction.

$$^{53}Cr + {}^{4}He \rightarrow \underline{\quad\quad} + 2n$$

The correct answer is: ^{55}Fe.

The mass number will be determined from the sum of the mass numbers of the reactants and products.

$53 + 4 = x + 2$ thus $x = 55$

$$^{53}_{24}\text{Fe} + {}^{4}_{2}\text{He} \rightarrow \underline{\quad} + 2\,{}^{1}_{0}\text{n}$$

The proton number will be determined from the sum of the proton numbers of the reactants and products.

$24 + 2 = x + 0$ thus $x = 26$

Fe has 26 protons. The isotope of Fe with a mass of 55 is ^{55}Fe.

INSIGHT: In all nuclear reactions the following rules are obeyed:
  1) The sum of the mass numbers of the reactants must equal the sum of the mass numbers of the products.
  2) The sum of the proton numbers of the reactants must equal the sum of the mass numbers of the products.

Kinetics of Radioactive Decay

7. **Tritium, ^{3}H, a radioactive isotope of hydrogen has a half-life of 12.26 y. If 2.0 g of ^{3}H were made, how much of it would be left 18.40 y later?**
   The correct answer is: 0.88 g.

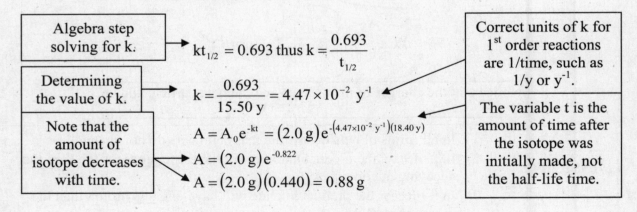

Algebra step solving for k.

$$kt_{1/2} = 0.693 \text{ thus } k = \frac{0.693}{t_{1/2}}$$

Determining the value of k.

$$k = \frac{0.693}{15.50 \text{ y}} = 4.47 \times 10^{-2} \text{ y}^{-1}$$

Note that the amount of isotope decreases with time.

$$A = A_0 e^{-kt} = (2.0 \text{ g})e^{-(4.47\times10^{-2} \text{ y}^{-1})(18.40\,\text{y})}$$
$$A = (2.0 \text{ g})e^{-0.822}$$
$$A = (2.0 \text{ g})(0.440) = 0.88 \text{ g}$$

Correct units of k for 1st order reactions are 1/time, such as 1/y or y^{-1}.

The variable t is the amount of time after the isotope was initially made, not the half-life time.

8. *A loaf of bread left in the Egyptian temple of Mentukotep II, an ancient pharaoh, has a ^{14}C content that is 61.6% that of living matter. How old is the loaf of bread? You may assume that the decrease in the ^{14}C content is entirely due to the radioactive decay of ^{14}C. The half-life of ^{14}C is 5730 y.*

The correct answer is: 4.00 x 10³ y.

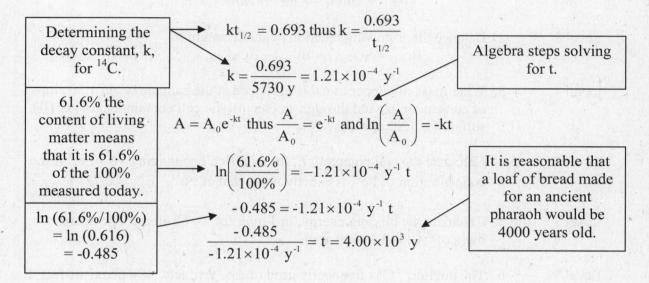

| Determining the decay constant, k, for ^{14}C. | $kt_{1/2} = 0.693$ thus $k = \dfrac{0.693}{t_{1/2}}$ |
| | $k = \dfrac{0.693}{5730 \text{ y}} = 1.21 \times 10^{-4} \text{ y}^{-1}$ |

Algebra steps solving for t.

$A = A_0 e^{-kt}$ thus $\dfrac{A}{A_0} = e^{-kt}$ and $\ln\left(\dfrac{A}{A_0}\right) = -kt$

61.6% the content of living matter means that it is 61.6% of the 100% measured today.

$\ln\left(\dfrac{61.6\%}{100\%}\right) = -1.21 \times 10^{-4} \text{ y}^{-1} \text{ t}$

It is reasonable that a loaf of bread made for an ancient pharaoh would be 4000 years old.

$\ln(61.6\%/100\%)$
$= \ln(0.616)$
$= -0.485$

$-0.485 = -1.21 \times 10^{-4} \text{ y}^{-1} \text{ t}$

$\dfrac{-0.485}{-1.21 \times 10^{-4} \text{ y}^{-1}} = t = 4.00 \times 10^3 \text{ y}$

Practice Test Six
Modules 19-20

Level 3   1. Balance the following reaction in both acidic and basic solution.
$$Fe^{2+} + MnO_4^- \rightarrow Fe^{3+} + Mn^{2+}$$

Level 3   2. Balance the following reaction in acidic solution.
$$Br_{2(l)} + SO_{2(g)} \rightarrow Br^-_{(aq)} + SO_4^{2-}{}_{(aq)}$$

Level 1   3. What mass of copper metal is produced at the cathode when 1.30 amps of current are passed through an electrolytic cell containing copper (II) sulfate for 72.0 minutes?

Level 1   4. Calculate the cell potential, E, for a half-cell containing Fe^{3+}/Fe^{2+} if the concentration of Fe^{2+} is exactly twice that of Fe^{3+}. $E^0_{cell} = 0.771\ V$

Level ?   5. Calculate the binding energy, in J/mol, for a ^{35}Cl atom. The actual mass of ^{35}Cl is 34.9689 amu.

Level ?   6. The nuclide ^{14}C is frequently used to date artifacts. If a given artifact has a ^{14}C content of 57.4% and the half-life of ^{14}C is 5730 years, then how old is the artifact? Assume that the decrease in ^{14}C content is entirely due to radioactive decay.

Level ?   7. Fill in the missing nuclide or decay particle in each nuclear reaction.

$$^{226}_{88}Ra \rightarrow \underline{\quad} + {}^{222}_{86}Rn$$

$$\underline{\quad} \rightarrow {}^{0}_{-1}\beta + {}^{37}_{18}Ar$$

$$^{15}_{7}N \rightarrow {}^{0}_{+1}\beta + \underline{\quad}$$

Math Review

Introduction

General chemistry classes require many basic mathematical skills. These include many which you were taught earlier in your academic career but may have forgotten from lack of use. This section will refresh your memory of the math skills necessary in the typical general chemistry course. The important topics in this section include:

1. the proper use of scientific notation
2. basic calculator skills, including entering numbers in scientific notation
3. the rounding of numbers
4. use of the quadratic equation
5. the Pythagorean theorem
6. rules of logarithms

Math Review Key Equations & Concepts

1. **The quadratic equation,** $x = \dfrac{-b \pm \sqrt{b^2 - 4ac}}{2a}$

   This equation is used to determine the solutions to quadratic equations, i.e. equations of the form $ax^2 + bx + c$. You will frequently encounter quadratic equations in equilibrium problems.

2. **They Pythagorean theorem,** $a^2 + b^2 = c^2$

   Used to determine the length of one side of a right triangle given the length of the other two sides of the triangle. This formula is frequently used in the section on the structure of solids to determine the edge length or the diagonal length of a cubic unit cell when calculating the atomic or ionic radius of an element.

3. $x = a^y$ **then** $y = \log_a x$

   $\log(x \cdot y) = \log x + \log y$

   $\log\left(\dfrac{x}{y}\right) = \log x - \log y$

   $\log(x^n) = n \log x$

   The first equation is the definition of logarithms. The other equations are basic rules of algebra using logarithms. These rules apply to logarithms of any base, including base e or natural logarithms, ln. These equations are frequently used in kinetics and thermodynamic expressions.

Scientific and Engineering Notation

In the physical and biological sciences it is frequently necessary to write numbers that are extremely large or small. It is not unusual for these numbers to have 20 or more digits beyond the decimal point. For the sake of simplicity and to save space when writing, a compact or shorthand method of writing these numbers must be employed. There are two possible but equivalent methods called either scientific or engineering notation. In both methods the insignificant digits that are placeholders between the decimal place and the significant figures are expressed as powers of ten. Significant digits are then multiplied by the appropriate powers of ten to give a number that is both mathematically correct and indicative of the correct number of significant figures to use in the problem. To be strictly correct, the significant figures should be between 1.000 and 9.999; however, this particular rule is frequently ignored. In fact, it must be ignored when adding numbers in scientific notation that have different powers of ten.

The only difference between scientific and engineering notation is how the powers of ten are written. Scientific notation uses the symbolism "x 10^y" whereas engineering notation uses the symbolism "Ey" or "ey". Engineering notation is frequently used in calculators and computers.

INSIGHT:

> *Positive powers of ten* indicate that the decimal place has been *moved to the left that number of spaces*.
>
> *Negative powers of ten* indicate that the decimal place has been *moved to the right that number of spaces*.

A few examples of both scientific and engineering notation are given in this table.

Number	Scientific Notation	Engineering Notation
10,000	1×10^4	1E4
100	1×10^2	1E2
1	1×10^0	1E0
0.01	1×10^{-2}	1E-2
0.000001	1×10^{-6}	1E-6
23,560	2.356×10^4	2.356E4
0.0000965	9.65×10^{-5}	9.65E-5

It is important for your success in chemistry that you understand how to use both of these methods of expressing very large or small numbers. Familiarize yourself with both methods.

Basic Calculator Skills

General Chemistry courses require calculations that are frequently performed on calculators. You do not need to purchase an expensive calculator for your course. Rather, you need a calculator that has some basic function keys. Common important functions to look for on a scientific calculator are: log and ln, antilogs or 10^x and e^x, ability to enter numbers in scientific or engineering notation, x^2, $1/x$, $\sqrt{}$ or multiple roots, like a cube or higher root.

More important than having an expensive calculator is knowledge of how to use your calculator. It is strongly recommended that you study the manual that comes with your calculator and learn the basic skills of entering numbers and understanding the answers that your calculator provides. For a typical general chemistry course there are three important calculator skills with which you should be proficient.

1) Entering Numbers in Scientific Notation

Get your calculator and enter into it the number 2.54×10^5. The correct sequence of strokes is: press 2, press the decimal button, press 5, press 4, and *then press either EE, EX, EXP or the appropriate exponential button on your calculator*. *Do not press x 10 before you press the exponential button!* This is a very common mistake and will cause your answer to be 10 times too large.

After you have entered 2.54×10^5 into your calculator, press the Enter or = button and look at the number display. If it displays 2.54E6 or 2.54×10^6, you have mistakenly entered the number. Correct your number entering method early in the course before it becomes a bad habit!

2) Taking Roots of Numbers and Entering Powers

Frequently we must take a square or cube root of a number to determine the correct answer to a problem. Most calculators have a square root button, $\sqrt{}$. To take a square root, simply enter the number into your calculator and press the $\sqrt{}$ button to get your answer. For example, take the square root of 72 (the answer is 8.49).

Some calculators have a $\sqrt[3]{}$ button as well. If your calculator does not have a $\sqrt[3]{}$ button, then you can use the y^x button to achieve the same result. To take a cube root, enter 1/3 or 0.333 as the power and the calculator will take a cube root for you. For example, enter $27^{0.333}$ into your calculator (the correct answer is 3.00). If you need a fourth root, enter ¼ which is 0.25 as the power, and so forth for higher roots.

3) Taking base 10 logs and natural or naperian logs, ln

Many of the functions in thermodynamics, equilibrium, and kinetics require the use of logarithms. All scientific calculators have log and ln buttons. To use them simply enter your number and press the button. For example, the log 1000 = 3.00, and the ln of 1000 = 6.91.

INSIGHT: A common mistake is taking the ln when the log is needed and vice versa. Be careful which logarithm you are calculating for the problem.

Rounding of Numbers

When determining the correct number of significant figures for a problem it is frequently necessary to round off an answer. Basically, if the number immediately after the last significant figure is a 4 or lower, round down. If it is a 6 or higher, round up. The confusion arrives when the determining number is a 5. If the following number is a 5 followed by a number greater than zero, round the number up. If the number after the 5 is a zero, then the textbook used in your course will have a rule based upon whether the following number is odd or even. You should use that rule to be consistent with your instructor. The following examples illustrate these ideas. In each case the final answer will contain three significant figures.

Initial Number	Rounded Number
3.67492	3.67
3.67623	3.68
3.67510	3.68
3.67502	Use your textbook rule.

Use of the Quadratic Equation

Equilibrium problems frequently require solutions of equations of the form $ax^2 + bx + c$. These are quadratic equations, and the two solutions can always be determined using this formula.

$$x = \frac{-b \pm \sqrt{b^2 - 4ac}}{2a}$$

For example, if the quadratic equation to be solved is $3x^2 + 12x - 6$, then a = 3, b = 12, and c = -6. The two solutions can be found in this fashion.

$$x = \frac{-b \pm \sqrt{b^2 - 4ac}}{2a}$$

$$x = \frac{-12 \pm \sqrt{12^2 - 4(3)(-6)}}{2(3)}$$

$$x = \frac{-12 \pm \sqrt{144 + 72}}{2(3)}$$

$$x = \frac{-12 \pm \sqrt{216}}{6}$$

$$x = \frac{-12 \pm 14.7}{6} = \frac{2.7}{6} \text{ and } \frac{-26.7}{6}$$

$$x = 0.45 \text{ and } -4.45$$

INSIGHT: Quadratic equations always have two solutions. In equilibrium problems, one of the solutions will not make physical sense. For example, it will give a negative concentration for the solutions or produce a concentration that is outside the possible ranges of solution concentrations. It is your responsibility as a student to choose the correct solution based on your knowledge of the problem.

The Pythagorean Theorem

In Module 13 we determined the radius of an atom in a cubic unit cell. Because the cell is cubic, a right triangle can always be formed using two of the sides and the face diagonal. The length of the face diagonal can be determined using the Pythagorean theorem. An example of the unit cell geometry and determining the face diagonal length is given below.

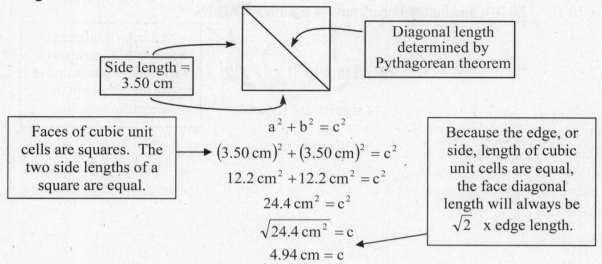

Side length = 3.50 cm

Diagonal length determined by Pythagorean theorem

Faces of cubic unit cells are squares. The two side lengths of a square are equal.

$$a^2 + b^2 = c^2$$
$$(3.50\,cm)^2 + (3.50\,cm)^2 = c^2$$
$$12.2\,cm^2 + 12.2\,cm^2 = c^2$$
$$24.4\,cm^2 = c^2$$
$$\sqrt{24.4\,cm^2} = c$$
$$4.94\,cm = c$$

Because the edge, or side, length of cubic unit cells are equal, the face diagonal length will always be $\sqrt{2}$ x edge length.

Rules of Logarithms

Logarithms are convenient methods of writing numbers that are exceptionally large or small and expressing functions that are exponential. They also have the convenience factor of making the multiplication and division of numbers written in scientific notation especially easy because in logarithmic form addition and subtraction of the numbers is all that is required. By definition, a logarithm is the number that the base must be raised to in order to produce the original number. For example, if the number we are working with is 1000 then 10, the base, must be cubed, raised to the 3rd power, to reproduce it. Mathematically, we are stating that $1000 = 10^3$, so the log $(1000) = 3$. There are three commonly used rules of logarithms that you must know. They are given below.

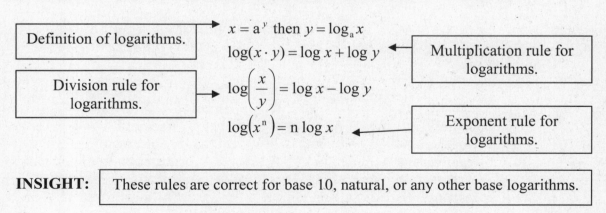

Definition of logarithms.

$$x = a^y \text{ then } y = \log_a x$$
$$\log(x \cdot y) = \log x + \log y$$

Multiplication rule for logarithms.

Division rule for logarithms.

$$\log\left(\frac{x}{y}\right) = \log x - \log y$$
$$\log(x^n) = n \log x$$

Exponent rule for logarithms.

INSIGHT: These rules are correct for base 10, natural, or any other base logarithms.

Significant Figures for Logarithms

There are 4 significant figures in the number 2.345 x 10^{12} (the 2, 3, 4, and 5). The power of 10 (the number 12) is not counted as significant. If we take the log of 2.345 x 10^{12} the number of significant figures must remain the same. The log of 2.345 x $10^{12} = 12.3701$.

What numbers indicate the exponents that are present in scientific notation? In logarithms, the numbers to the left of the decimal place (the characteristic) are insignificant and the ones to the right (the mantissa) are significant. Thus, the log (2.345 x 10^{12}) = 12.3701, and both numbers have 4 significant figures.

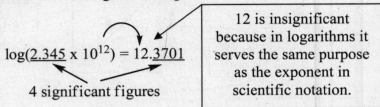

$$\log(\underline{2.345} \times 10^{12}) = 12.\underline{3701}$$

4 significant figures

12 is insignificant because in logarithms it serves the same purpose as the exponent in scientific notation.

Practice Test One Solutions

1. 5.31×10^6 dm^3

2. 1.468×10^5

3. 5.12 g

4. 4.65×10^{24} atoms of N

5. 1.817 mol Ca$_3$(AsO$_4$)$_2$

6. a) 2 iron (III) ions
   b) 3 sulfate ions
   c) 0 sulfide ions
   d) 0 oxide ions

7. a) phosphorus pentachloride
   b) ammonium sulfate
   c) lithium nitrate
   d) potassium dihydrogen hypobromite
   e) xenon tetrafluoride

8. a) SF$_6$
   b) HCN
   c) Cu(OH)Cl
   d) MgBr$_2$
   e) HClO

Practice Test Two Solutions

1. 22.8 g oxygen

2. **1** P$_4$O$_{10}$ + **6** H$_2$O → **4** H$_3$PO$_4$

3. AgNO$_3$ is the limiting reagent, and 13.73 g of Ca(NO$_3$)$_2$ is the theoretical yield. The percent yield is 78.10 g.

4. 6.81 mol HCl

5. The required volume is 441 mL.

6. The final volume is 130. mL.

7. $2H_2O \rightarrow 2H_2 + O_2$ decomposition and oxidation/reduction
$H_2 + Cl_2 \rightarrow 2HCl$ combination and oxidation/reduction
$AlCl_3 + 3AgNO_3 \rightarrow 3AgCl + Al(NO_3)_3$ metathesis and precipitation

8. Total ionic equation: $2HClO_3 + Sr^{2+} + 2OH^- \rightarrow Sr^{2+} + 2ClO_3^- + 2H_2O$
Net ionic equation: $HClO_3 + OH^- \rightarrow ClO_3^- + H_2O$

Practice Test Three Solutions

1. $1s^2 2s^2 2p^6 3s^2 3p^6 4s^2 3d^8$ or $[Ar]\, 4s^2 3d^8$
   a) $n = 3$
   b) 6 paired electrons
   c) 2 unpaired electrons
   d) $\ell = 2$

2. a) 10
   b) 2
   c) 14
   d) 6
   e) 2

3. Oxygen is the most electronegative of the atoms and N has the highest first ionization energy.

4. Cl releases the most energy upon accepting an electron. Its electron affinity is very negative.

5. Atomic radii increase down a group because of the increase in principle quantum number. Electrons are farther away from the nucleus. From left to right across a period, the principle quantum number is the same; however, effective nuclear charge increases due to decreased shielding. The increase in effective nuclear charge pulls the electrons in closer to the nucleus, resulting in a smaller radius.

6. MgO is an ionic compound (metal and nonmetal). There are two ions, Mg^{2+} and O^{2-}.

$$[Mg]^{2+}\ [:\ddot{O}:]^{2-}$$

7. SF_4 is a covalent compound (two nonmetals). There are no ions present. In the Lewis structure, S is the central atom with four single bonds to F atoms. There is also a lone pair on the central S atom.

8. XeF_4 octahedral electronic geometry; square planar molecular geometry
   I_3^- trigonal bipyramidal electronic geometry; linear molecular geometry
   CO_2 linear electronic and molecular geometry
   C_2H_2 linear electronic and molecular geometry about both C atoms

9. All of the molecules in question 8 are non polar.

10. XeF_4 sp^3d^2
     I_3^- sp^3d
     CO_2 sp
     C_2H_2 sp for both C atoms

Practice Test Four Solutions

1. Strong acids: HNO_3, HCl, HI, HBr, H_2SO_4, $HClO_4$ ($HClO_3$ would also be correct.)
   Strong bases: LiOH, NaOH, KOH, RbOH, CsOH, $Ca(OH)_2$, $Sr(OH)_2$, $Ba(OH)_2$

2. The true statements are a), c), and e).

3. a) CH_4 London dispersion
    b) CH_2Cl_2 diple-dipole
    c) CH_3COOH hydrogen bonding
    d) HF   hydrogen bonding
    e) PCl_3 dipole-dipole

4. $CaO > CH_3COOH > CH_2Br_2 > CCl_4$

5. $V = 49.6$ L

6. There are two atoms per unit cell, so it is a body-centered cubic unit cell.

7. % w/w = 21.2%
   $X_{H_3PO_4} = 0.0472$

8. 34. g/mol

9. 7.03×10^5 J

10. $\Delta H^0_{rxn} = -1516.8$ J
   The negative sign indicates that the reaction is exothermic.

Practice Test Five Solutions

1. rate = $k[A]^2[C]$

2. The reaction is first order (from the units of the rate constant); $[A] = 0.160$ M

3. $t = 7.92 \times 10^3$ s or 132 min

4. $E_a = 3.00 \times 10^4$ J/mol

5. $K_c = 8.0 \times 10^{-12}$

6. $[PCl_3] = 0.071$ M

7. Since the reaction is endothermic, heat is a reactant. Increasing the temperature is the equivalent of adding reactant. Addition of a reactant shifts equilibrium toward the products.

8. pH = 7.26

9. pOH = 6.13; pH = 7.87

10. pH = 3.89

Practice Test Six Solutions

1. Acidic solution: $5Fe^{2+} + MnO_4^- + 8H^+ \rightarrow 5Fe^{3+} + Mn^{2+} + 4H_2O$
   Basic solution: $5Fe^{2+} + MnO_4^- + 8H_2O \rightarrow 5Fe^{3+} + Mn^{2+} + 4H_2O + 8OH^-$

2. $Br_2 + 2H_2O + SO_2 \rightarrow 2Br^- + SO_4^{2-} + 4H^+$

3. 1.85 g $Cu_{(s)}$

4. $E = 0.753$ V

5. BE = 2.8827×10^{16} J/mol atoms

6. 4.59×10^3 y

7. $^4_2 He$
   $^{37}_{17} Cl$
   $^{15}_6 C$